# PID Tuning

# PID Tuning

## A Modern Approach via the Weighted Sensitivity Problem

Salvador Alcántara Cano

Ramon Vilanova Arbós

Carles Pedret i Ferré

CRC Press

Taylor & Francis Group

Boca Raton London New York

CRC Press is an imprint of the
Taylor & Francis Group, an **informa** business

# Contents

# Foreword

Control engineering is one of the disciplines with a more profound impact on the development of our society. This is supported with new theories, controllers, actuators, sensors, industrial processes, computer methods, applications, and design philosophies to tackle the challenges in front of us. As a today's out-of-the-box example, it's interesting to see how reconcilliation-engine techniques (feedback!) have become widespread within the IT world: React.js for UI, Kubernetes for container fleets, Terraform for managing cloud resources, and there is probably more to come.

Even if they date back to 1939, when the Taylor and Foxboro instrument companies introduced the first PID controllers, present-day controllers are still based on those original *proportional*, *integral*, and *derivative* modes and constitute the workhorse of modern process control systems. Although advanced control techniques can provide significant improvements, a PID controller that is properly designed and tuned has proven to be satisfactory for the vast majority of industrial control loops. In fact, as confirmed by a recent survey published in *Control System Magazine* [CSM, Feb 2017], PID controllers continue to appear among the top control technologies in industry.

In my more than 20 years of control theory research career, I have been exposed to a variety of control systems, from papermaking, refining, power generation, power grid, motors, and instruments to ships, satellites, aircraft, and robots, as well as in-depth exchanges with many engineers. Based on these experiences, I have a persistent preference for simple and effective design methods and simple and effective controllers. The authors of this book and I have the same preference. The book studies the design problem with emphasis on the simple and effective PID controller. The analytical method proposed by the book yields good performance, is motivated by a rigorous framework, and simplifies the design process as much as possible. This is the concern of engineers.

This book on *PID Tuning: A Modern Approach via the Weighted Sensitivity Problem*, by Professors Alcántara, Vilanova, and Pedret covers most of the essential aspects, not only regarding the three-term controller design but also about feedback control at large. The problem considered in this book is embodied with a set of modern control concepts usually found only in pure theoretical works. Here it is shown how they can also fit perfectly into the industrial realm. This book will be a valuable resource for both engineering students and practitioners searching for guidance and insights into analytical procedures and the tuning task itself.

**Weidong Zhang**
*Department of Automation, Shanghai Jiao Tong University*
*April 13th, 2020*

# Preface

This book serves as a practical introduction to optimal and robust control for those with a basic working knowledge of signals and (control) systems. Its ultimate goal is to help bridge the gap between classical and more advanced control theories. The word *modern* in the title refers to the approaches and developments that have taken place in control theory since the 1980s, and that require a higher level of formalism and rigor compared with the classical methods with which the reader is already familiar.

To reach its objective, only single-input/single-output, linear time-invariant, and continuous-time systems are considered, making the book rather accessible to a wide audience. The *weighted sensitivity problem* is adopted as the vehicle for the main ideas, which are applied to PID control due to the paramount importance of this simple controller in industry. Overall, the contents are particularly well-suited for final-year undergraduate, as well as graduate students. Control practitioners, novice and professional alike, might find it interesting and useful, too.

The outline is as follows:

- Part I goes through the design of PID controllers for stable plants using a model matching approach. This gives rise to preliminary tuning rules that will be improved later on in the book, but most importantly introduces the *model-matching problem* along with other important concepts such as the $\mathcal{H}_\infty$ norm, *nominal* and *robust stability*, *smooth/tight* control, and the *robustness/performance* and *servo/regulation* trade-offs.

- Part II revolves around the main theme of the book: the *weighted sensitivity problem*, which has proved to be a convenient one for designing feedback compensators in general. The first step is to guide the reader on the weight selection. Then, both the $\mathcal{H}_\infty$ and the $\mathcal{H}_2$ norms are considered separately. For the former, it is shown how using possibly unstable plants results into a plausible trick for avoiding any notion of *coprime factorization*, therefore unifying (and greatly facilitating) the analytical procedure. For the latter, the connection with the well-known two-step *internal model control* procedure is fully addressed. In particular, the proposed weight is interpreted in terms of IMC filters.

- Part III retakes the subject of PID tuning based on the methods presented in Part II. Because the particular norm used does not play a key role, only the $\mathcal{H}_\infty$ norm is considered in the remainder of the book. This part finally derives improved tuning expressions for the most common first- and second-order models used in practice. The proposed tuning rules generalize many existing ones given in the literature, stressing their unifying character. More precisely, what makes them unique is that they tackle both stable and unstable plants, while at the same time explicitly considering the robustness/performance and servo/regulation trade-offs by means of two parameters with clear engineering meaning. Finally, this part concludes by showing how to tune these design parameters for *balanced* operation.

In summary, this book concisely covers the topics required for transitioning from classical to more advanced control theories. This is done without abandoning the ubiquitous PID controller along the way. By sticking to this simple realization, we find that it is easier to provide insight into the tuning procedure, which is valuable from both pedagogical and practical standpoints.

## Acknowledgements

The authors would like to acknowledge all the people who have contributed to this book in one way or another, such as S. Skogestad, W. Zhang, S. Dormido, A. Visioli, O. Arrieta, F. Padula, and V.M. Alfaro, just to name a few–in no relevant order. Special thanks go to the CRC Press publisher for Engineering, Gagandeep Singh, as well as Annie Sophia and their team at Nova Techset, for their help during the preparation of the manuscript. Partial support for the research that originated the results presented in this book was provided by the Spanish Ministry of Economy and Competitivity through grant DPI-2016-77271-R.

The first author, "Salva", is also very thankful to many people he has met since he left university: C. Gabaldón, A. Waalkens, F. Sempere, D. Hidalgo, D. Bravo, P. San Valero, D. Vílchez, A. Visser (at PAS/UV, within the Next Air Biotreat project), J. Ruiz, R. Barrio, X. Tornero, Ó. Dieste de Sus, V. Ferré, A. Linares (at Sipro), Josep Sanjuas, Valentín, Jordi, Karol, Gerard, Miguel, Ruben, Johan, José (at Talaia), and Nanook at Auvik.

# Authors

**Salvador Alcántara Cano** graduated in Computer Science & Engineering and then obtained the MSc and PhD degrees in Systems Engineering & Automation, all from Universitat Autònoma de Barcelona, in 2005, 2008, and 2011, respectively. During his short-lived research career, he focused on PID control and the analytical derivation of simple tuning rules guided by robust and optimal principles. He also made two research appointments with Professors Weidong Zhang and Sigurd Skogestad, almost completed a degree in Mathematics, and held a Marie Curie postdoctoral position in the Netherlands. Back in Barcelona, "Salva" worked as an automation & control practitioner for one more year, before definitively shifting his career into software development. Apart from programming and DevOps in general, his current interests include Stream Processing, Machine Learning, and Functional Programming & Category Theory.

**Ramon Vilanova Arbós** graduated from the Universitat Autònoma de Barcelona (1991), obtaining the title of Doctor through the same university (1996). At present, he's Full Professor of Automatic Control and Systems Engineering at the School of Engineering of the Universitat Autònoma de Barcelona where he develops educational task-teaching subjects of Signals and Systems, Automatic Control, and Technology of Automated Systems. His research interests include methods of tuning of PID regulators, systems with uncertainty, analysis of control systems with several degrees of freedom, applications to environmental systems, and development of methodologies for the design of machine-man interfaces. He is an author of several book chapters and has more than 100 publications in international congresses/journals. He is a member of IFAC and IEEE-IES. He's also a member of the Technical Committee on Factory Automation.

**Carles Pedret i Ferré** was born in Tarragona, Spain, on January 29, 1972. He received the BSc degree in Electronic Engineering and the PhD degree in System Engineering and Automation from the Universitat Autònoma de Barcelona, in 1997 and 2003, respectively. He is Associate Professor at the Department of Telecommunications and System Engineering of the Universitat Autònoma de Barcelona. His research interest are in uncertain systems, time-delay systems, and PID control.

# 1

## Introduction

The aim of this introduction is to guide the reader through the content of subsequent chapters, offering a high-level presentation of the underlying concepts and ideas that appear all throught the book. The discussion that follows is based on the unity feedback, linear time invariant (LTI), and single-input-single-output (SISO) continuous-time system of Figure 1.1, where $P$ is the plant and $K$ the feedback controller to be designed. The output signals $y, u$ represent, respectively, the plant output and the control action. Two exogenous inputs to the system are considered: $d$ and $r$. Here, $d$ represents a disturbance affecting the plant input, and the term *regulator mode* refers to the case when this is the main exogenous input.

### 1.1   Servo, regulation, and stability

The term *servo mode* refers to the case when the set-point change $r$ is the main concern. Although the reference tracking can be improved by using a two-degree-of-freedom (2DOF) controller [72, 20], there will always be some unmeasured disturbance directly affecting the plant output, which may be represented as an unmeasured signal $r$ (in this case, the plant output would be the control error $e = r - y$). In summary, there is a fundamental trade-off between the regulator (input disturbance) and servo (output disturbance) modes. The closed-loop mapping for the system in Figure 1.1 is given by

$$\begin{bmatrix} y \\ u \end{bmatrix} = \begin{bmatrix} T & SP \\ KS & S \end{bmatrix} \begin{bmatrix} r \\ d \end{bmatrix} \doteq H(P,K) \begin{bmatrix} r \\ d \end{bmatrix} \tag{1.1}$$

where $S \doteq \frac{1}{1+PK}$ and $T \doteq \frac{PK}{1+PK}$ denote the sensitivity and complementary sensitivity functions [72], respectively (note that $S + T = 1$). In terms of the performance for the regulator and servo modes, the closed-loop effect of disturbance and set-point changes on the output error is given by

$$y - r = -e = -Sr + SPd \tag{1.2}$$

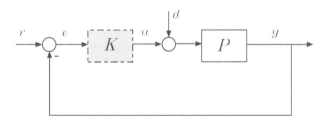

**FIGURE 1.1**
Conventional feedback configuration.

The most basic requirement for the controller $K$ is *internal stability*, which means that all the relations in $H(P, K)$ are stable. The set of all internally stabilizing feedback controllers will be hereafter denoted by $\mathcal{C}$. At this point, it is also convenient to introduce a special notation for the set of stable transfer functions, or $\mathcal{RH}_\infty$ for short. Many times, arguments in signals and transfer functions will be dropped for simplicity, i.e., we will write $y$ instead of $y(t)$ or $y(s)$, and $P$ instead of $P(s)$. We also note here that $P$ will normally be used to refer both to the real plant and the model of it. Depending on the context, however, we will also use the notation $\tilde{P}$ for the real (uncertain) plant in order to distinguish it from the model.

As this book concentrates on the design of feedback controllers of proportional, integral, and derivative (PID) type that will be designed from a modern control perspective, next section will present in a succinct way the PID controller. Also, as the weighted sensitivity approach to PID design has received the influence of internal model and $\mathcal{H}_\infty$ control, these two paradigms are also briefly outlined next in order to better present the book outline and the road to the obtention of tuning rules for balanced operation.

## 1.2    Industrial PID control

PID controllers have been around in the process industry for more than seven decades [51]. In spite of their old existence, current surveys estimate that the great majority of the controllers are (still) of PID type. For example, in [37], the process control state of the art in Japan is surveyed, and it is reported that the ratio of applications of PID control, conventional advanced control, and MPC is 100:10:1. The main reasons for the great success of PID controllers are that they are well-performing in many applications, and that they are easy to understand and implement. Furthermore, today's technology provides additional features like automatic tuning or gain scheduling [20]. Add to this a long history of proven operations, and it is not difficult to understand the decision.

Originally, the PID algorithm was conceived as the combination of three basic control actions, hence its name, so that the control law can be ideally expressed as

$$u(t) = K_c \left( e(t) + \frac{1}{T_i} \int_0^t e(\tau)d\tau + T_d \frac{de(t)}{dt} \right) \tag{1.3}$$

where the tuning parameters $K_c, T_i, T_d$ are known as the proportional gain, integral, and derivative times, respectively. Accordingly, the ideal PID law is based on the present ($e(t)$), past ($\int_0^t e(\tau)d\tau$), and estimated future ($\dot{e}(t)$) error information. PID controllers can also be understood in terms of lead-lag compensation. Even for such a simple strategy, it is not easy to find good settings for $K_c, T_i, T_d$ without a systematic procedure [57, 52]. In this regard, a visit to a process plant will usually show poorly tuned PID controllers [69].

Because pure derivative action cannot be implemented, the following *commercial* PID transfer function is usually considered:

$$K = K_c \left( 1 + \frac{1}{sT_i} + \frac{sT_d}{1 + sT_d/N} \right) \tag{1.4}$$

In the (noninteractive) industrial ISA form (1.4), $N$ is the derivative filter parameter. Although $N$ is normally fixed by the manufacturer (or restricted to a limited range), the advantages of considering $N$ an extra tuning parameter have been stressed in different works [46, 36, 44]. Adhering to this recommendation, $N$ will be considered tunable. In fact, disregarding $N$ at the design stage is, in part, responsible for the myth that derivation action does not work (it is common to find PID controllers acting as PI controllers, namely $T_d = 0$, to avoid noise sensitivity problems). As reported in [39, 41], improved filtering of PID controllers has a great potential to show industry the benefits of derivative action.

From a modern perspective, a PID controller is simply a controller of up to second order containing an integrator:

$$K = \frac{c_1 s^2 + c_2 s + c_3}{s(d_1 s + 1)} \tag{1.5}$$

where $c_1, c_2, c_3, d_1$ are positive real constants. Considering the general second-order form (1.5) is quite natural [36] and helps to avoid the computational oddities that may exist in the PID algorithms of common vendors. For example, $N$ and $T_d$ may be negative when going from (1.5) to (1.4). This problem is avoided if the *practical* output-filtered form (1.6) introduced in [49] is used for implementation:

$$K = K_c \left( 1 + \frac{1}{sT_i} + sT_d \right) \frac{1}{T_F s + 1} \tag{1.6}$$

where $T_F$ is a fourth tuning parameter. A general reformulation of the PID controller along the lines of (1.5), but including up to five tuning parameters, can be found in [39].

Although it is surprising that such a simple structure (1.5) works so well, the simplicity entails limitations too, implying that in some applications better performance can be obtained using more sophisticated strategies. For example, for processes with long time delays it is better to combine the PID controller with a dead-time compensator like the Smith predictor [30]. Another situation where the the PID controller (alone) is not recommended is for oscillatory processes [20]. Even in such cases, a PID controller properly augmented with an additional low-pass filter may yield acceptable results in practice [39]. Consequently, although simple candidates to replace the PID compensator have appeared in the literature [56, 52], it seems that PID control has a future yet. Definitely, it still constitutes an active research field, including topics such as (automatic) tuning methodologies [44], adaptive and robust control [60], stabilizing parameters [33], fractional control [54], or event-based control [76].

## 1.3 Internal model and $\mathcal{H}_\infty$ control

The design approach presented in this book has received the influence of internal model and $\mathcal{H}_\infty$ control. These two paradigms are briefly outlined next. A design method that puts them together will also be reviewed.

### 1.3.1 Internal model control

Let us start factoring the plant as $P = P_a P_m$, where $P_a \in \mathcal{RH}_\infty$ is all-pass and $P_m$ is minimum-phase (MP). As reported in [72, 25], the broad objective of the internal model control (IMC) procedure [49] is to specify the closed-loop relation $T_{yr} = T = P_a f$, where $f$ is the so-called IMC filter. Assuming that $P$ has $k$ unstable poles, the filter is chosen as follows:

$$f(s) = \frac{\sum_{i=1}^{k} a_i s^i + 1}{(\lambda s + 1)^{n+k}} \tag{1.7}$$

The purpose of $f$ is twofold: first, to ensure the properness of the controller and the internal stability requirement (to this double aim, $n$ must be equal or greater than the relative degree of $P$, whereas the $a_1, \ldots, a_k$ coefficients impose $S = 0$ at the $k$ unstable poles of $P$). Second, the $\lambda$ parameter is used to find a compromise between robustness and performance. Roughly speaking, the choice $T = P_a f$ is motivated by $\mathcal{H}_2$ optimization[1], in such a way that when $\lambda \to 0$ the closed-loop minimizes $\|S\frac{1}{s}\|_2$. By using Parseval's theorem and the fact that $S = T_{er}$, as it is clear from (1.2), minimizing $\|S\frac{1}{s}\|_2$ is equivalent to minimizing the Integrated square error (ISE) over time due to a set-point change. Starting with a small value of $\lambda$, optimality can then be sacrificed for the sake of robustness (this process is referred to as *detuning*). In general, the larger the value of $\lambda$, the more robust (and slow) the resulting system. For example, in the stable plant case—i.e., $k = 0$ in (1.7)—it is readily seen that $\lambda$ is closely related to the closed-loop bandwidth since $|T| = |P_a f| = |f| = \frac{1}{|(\lambda j\omega + 1)^n|}$.

The main advantages of IMC are its simplicity and analytical character, while some of its drawbacks are listed below[2]:

- For stable plants, the poles of $P$ are cancelled by the zeros of the controller $K$. This yields good results in terms of set-point tracking but results in sluggish disturbance attenuation when $P$ has slow/integrating poles [23, 34, 69, 64].

- For unstable plants, the pole-zero pattern of (1.7) can lead to large peaks on the sensitivity functions, which in turn means poor robustness and large overshoots in the transient response [22].

- In general, poor servo/regulation performance compromise is obtained [69].

### 1.3.2   $\mathcal{H}_\infty$ control

Modern $\mathcal{H}_\infty$ control theory [72] is based on the general feedback setup depicted in Figure 1.2, composed of the generalized plant $G$ and the feedback controller $K$. Once the problem has been posed in this form, the optimization process aims at finding a controller $K$, which makes the feedback system in Figure 1.2 stable, and minimizes the $\mathcal{H}_\infty$ norm[3] (sometimes referred to as *min-max* or supremum norm) of the closed-loop relation from $w$ to $z$. Mathematically, the synthesis problem can be expressed as

$$\min_{K \in \mathcal{C}} \|\mathcal{N}\|_\infty = \min_{K \in \mathcal{C}} \|\mathcal{F}_l(G, K)\|_\infty \tag{1.8}$$

---

[1] For a SISO (strictly proper) LTI system $G(s)$, the $\mathcal{H}_2$ norm is defined as

$$\|G(s)\|_2 \doteq \left( \frac{1}{2\pi} \int_{-\infty}^{\infty} |G(j\omega)|^2 d\omega \right)^{1/2}$$

Parseval's theorem says that

$$\|G(s)\|_2^2 = \|g(t)\|_2^2 \doteq \int_0^\infty |g(t)|^2 dt$$

where $g(t) = \mathcal{L}^{-1}(G(s))$ denotes the impulse response of $G$. Therefore, $\|G(s)\|_2$ can be interpreted in terms of the integrated square error of the impulse response. For more performance interpretations (both deterministic and stochastic), and a definition covering the multivariable case, consult [72].

[2] For an exhaustive list of the IMC shortcomings, consult [25, Section 3]. Here, only the points relevant to the book are considered.

[3] For a (proper) LTI system $G(s)$, the $\mathcal{H}_\infty$ norm is defined as

$$\|G(s)\|_\infty \doteq \max_\omega \bar{\sigma}(G(j\omega))$$

where $\bar{\sigma}$ denotes the largest singular value (induced 2-norm). In the SISO case, $\bar{\sigma}(G(j\omega)) = |G(j\omega)|$. Thus, the $\mathcal{H}_\infty$ norm is simply the peak of the transfer function magnitude. By introducing weights, the $\mathcal{H}_\infty$ norm can be interpreted as the magnitude of some closed-loop transfer function(s) relative to a specified upper bound [72].

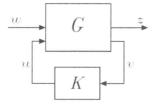

**FIGURE 1.2**
Generalized control setup.

where

$$\mathcal{N} = \mathcal{F}_l(G, K) \doteq G_{11} + G_{12}K(I - G_{22}K)^{-1}G_{21} = T_{zw} \tag{1.9}$$

An important feature of the $\mathcal{H}_\infty$ norm is that it not only captures performance but also robustness objectives. More concretely, uncertainty (usually denoted by $\Delta$) can be explicitly considered in the general control configuration; see Figure 1.3. Then, if $\|\mathcal{N}_{11}\|_\infty < 1$, the Small-Gain theorem guarantees that the closed-loop remains stable for every (normalized) perturbation such that $\|\Delta\|_\infty < 1$ (this is referred to as *robust stability*). Thus, using the $\mathcal{H}_\infty$ norm, it is possible to deal with performance and robustness issues simultaneously by means of the so-called mixed sensitivity problems. From this point of view, the basic idea behind the $\mathcal{H}_\infty$ design methodology is to press down the peaks of several closed-loop transfer functions; some may be related to performance and some others to robustness. The main difficulty with the $\mathcal{H}_\infty$ methodology is that the designer has to select suitable frequency weights included in $G$ (this point will be exemplified shortly later), which may require considerable trial and error. Furthermore, for stacked problems involving three or more closed-loop transfer functions, the shaping becomes considerably difficult for the designer [72].

### 1.3.3 Blending internal model and $\mathcal{H}_\infty$ control

In [25], a systematic $\mathcal{H}_\infty$ procedure to generalize IMC is presented. The idea is to avoid the limitations of the IMC design method when these are present but retain its desirable features and simplicity when these shortcomings are absent. To achieve this goal, the following problem is posed to be solved numerically:

$$
\begin{aligned}
\rho &= \min_{K \in \mathcal{C}} \|\mathcal{N}\|_\infty \\
&= \min_{K \in \mathcal{C}} \left\| \mathcal{F}_l \left( \begin{bmatrix} -P_a f & \epsilon_2 P & P \\ 0 & \epsilon_1\epsilon_2 & \epsilon_1 \\ 1 & -\epsilon_2 P & -P \end{bmatrix}, K \right) \right\|_\infty \\
&= \min_{K \in \mathcal{C}} \left\| \begin{array}{cc} T - P_a f & \epsilon_2 SP \\ \epsilon_1 KS & \epsilon_1\epsilon_2 S \end{array} \right\|_\infty
\end{aligned}
\tag{1.10}
$$

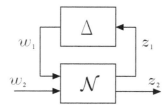

**FIGURE 1.3**
Generalized control setup including uncertainty.

where $\epsilon_1$ and $\epsilon_2$ are stable, MP and proper weighting functions. The basic philosophy is to minimize the closeness between the input-to-output relation and a specified reference model, which is set as $P_a f$ along the lines (but with more flexibility) of the standard IMC. At the same time, the (1,2) term of (1.10) limits the size of $SP = T_{yd}$, whereas the (2,1) term limits the size of $KS = T_{ur}$. The index in (1.10) automatically guarantees that

$$|T(j\omega) - P_a f(j\omega)| \leqslant \rho \; \forall \omega, \tag{1.11}$$

$$|SP(j\omega)| \leqslant \rho/|\epsilon_2(j\omega)| \; \forall \omega, \tag{1.12}$$

$$|KS(j\omega)| \leqslant \rho/|\epsilon_1(j\omega)| \; \forall \omega \tag{1.13}$$

Now, if the design specifications are written as $\|T - P_a f\|_\infty \leqslant \alpha$, $|SP(j\omega)| \leqslant \beta_p^i \; \forall \omega \in [w_1^i, w_2^i]$ and $|KS(j\omega)| \leqslant \beta_k^i \; \forall \omega \in [\omega_3^i, \omega_4^i]$, where $\alpha, \beta_p^i, \beta_k^i, w_1^i, w_2^i, w_3^i$ and $w_4^i$ are positive real numbers representing the closed-loop objectives, $\epsilon_1$ and $\epsilon_2$, can be chosen as

$$|\epsilon_1(j\omega)| \geqslant \alpha/\beta_k^i \; \forall \omega \in [\omega_3^i, \omega_4^i] \quad \text{and} \quad |\epsilon_2(j\omega)| \geqslant \alpha/\beta_p^i \; \forall \omega \in [w_1^i, w_2^i] \tag{1.14}$$

Then, if $\rho \leqslant \alpha$, the design specifications are certainly met[4]. Note that by selecting $\epsilon_1 = \epsilon_2 = 0$ (corresponding to $\beta_k^i, \beta_p^i \to \infty$) and $f$ as in (1.7), the design reduces, essentially, to the original IMC procedure.

The revised design method has great versatility, blending IMC and $\mathcal{H}_\infty$ ideas elegantly[5]. In exchange, the resulting procedure inevitably loses part of the IMC simplicity (even if $f, \epsilon_1, \epsilon_2$ can be chosen in a systematic way) and its analytical character. Without some caution, this may translate into design pitfalls as noted in [42]. Another disadvantage of the $\mathcal{H}_\infty$ machinery is that it usually gives high-order controllers, requiring the use of model order reduction techniques [72].

As it was pointed out in Section 1.1, some basic problems with IMC are related to servo/regulation issues. In this book, the design methodology will share the analytical character of IMC and much of its simplicity. This will be achieved by considering a simple weighted sensitivity formulation. Apart from considering the inherent compromise between robustness and performance, extra design parameters will be finally introduced into the weight to deal with the trade-off between the servo and regulatory performance.

### 1.3.4   Vilanova's (2008) design for robust PID tuning revisited

Let us consider the following problem

$$\min_{K \in \mathcal{C}} \left\| W(T_d - T) \right\|_\infty \tag{1.15}$$

where $T_d$ represents the desired complementary sensitivity (the input-to-output response: $T_{yr}$) and $W$ is a frequency weight (see Figures 1.4 and 1.5). The idea in [81] is to use simple settings in order to obtain a PID controller:

- $T_d = \frac{1}{T_M s + 1}$, where the $T_M$ parameter specifies the desired *speed of response*.

- $W = \frac{zs+1}{s}$. The integrator forces integral action by requiring perfect matching between $T$ and $T_d$ at zero frequency. The $z$ parameter is used to adjust the robustness margins: the larger the value of $z$, the more robust the resulting system.

---

[4]Notice that if $\rho > \alpha$ one cannot conclude anything about achieving the performance objectives.

[5] The reader may be aware of other techniques with strong links to the size of $H(P, K)$. One is the $H_\infty$ loop shaping method [47, 72], which has points in common with the design philosophy in [25]. Note, however, that the frequency cost functions $\epsilon_1$ and $\epsilon_2$ capture the design objectives differently.

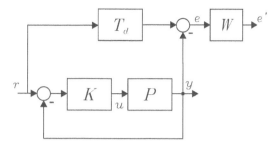

**FIGURE 1.4**
Diagram for problem (1.15). Here, $e$ denotes the error between the desired and actual output.

In addition, many processes have rather simple dynamics and are often modelled using low-order models of the form

$$P = K_g \frac{e^{-sh}}{\tau s + 1} \tag{1.16}$$

called first-order plus time delay or just FOPTD systems, where $K_g, h, \tau$ are, respectively, the gain, the (apparent) delay, and the time constant of the process. These models can be obtained easily through open-loop and closed-loop step response tests [66]. Alternatively, one can start from an accurate description of the process and apply then some model reduction technique. In this regard, the *half-rule* in [69] provides a simple analytic approach. Supported by these considerations, a stable FOPTD model is used in [81]. However, for derivation purposes, the time delay in (1.16) is approximated using a first-order Taylor expansion:

$$P = K_g \frac{-sh + 1}{\tau s + 1}$$

Note that the problem at hand could now be posed in terms of the general control setup (see Figure 1.2) to be solved numerically for particular values of the tuning parameters. For such a simple problem, however, the solution can be obtained analytically. First, note that for stable plants all stabilizing controllers as can be expressed as

$$K = \frac{Q}{1 - PQ} \tag{1.17}$$

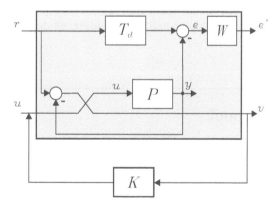

**FIGURE 1.5**
Problem (1.15) rearranged in the general control form.

Then, in terms of $Q$, $T = PQ$, and problem (1.15) gets equivalent to

$$\min_{Q \in \mathcal{RH}_\infty} \left\| W(T_d - PQ) \right\|_\infty \tag{1.18}$$

which is a model-matching problem with $T_1 = WT_d, T_2 = WP$. The solution to (1.18) follows from [81] and the resulting feedback compensator obtained turns out to be a PID controller:

$$K = \frac{1}{K_g(\rho + T_M)} \frac{(1 + \tau s)(1 + \chi s)}{s(1 + \frac{zT_M + h\chi}{\rho + T_M} s)} \tag{1.19}$$

where

$$\begin{aligned}
K_c &= \frac{T_i}{K_g(\rho + T_M)} \\
T_i &= \tau + \chi - T_M \frac{\rho + z}{\rho + T_M} \\
\frac{T_d}{N} &= T_M \frac{\rho + z}{\rho + T_M} \\
N + 1 &= \frac{\tau}{T_i} \frac{\chi}{T_M} \frac{\rho + T_M}{\rho + z}
\end{aligned} \tag{1.20}$$

and

$$\rho - \frac{h + z}{h + T_M} h \qquad \chi = h + z - \rho \tag{1.21}$$

The tuning parameters $T_m, z$ can be fixed for auto-tuning purposes:

$$T_M = 2h \qquad z = \sqrt{2}h \tag{1.22}$$

With these values for $z, T_M$, which are chosen according to robustness considerations, the tuning rule (in ISA-PID form) can be expressed as

$$\begin{aligned}
K_c &= \frac{T_i}{K_g h 2.65} \\
T_i &= \tau + 0.03h \\
\frac{T_d}{N} &= 1.72h \\
N + 1 &= \frac{\tau}{T_i}
\end{aligned} \tag{1.23}$$

The tuning rule in [81] has been found to work well in practice, exhibiting a robust behaviour in the face of process abnormalities including model mismatch, valve stiction, and sensor noise [14].

## 1.4 Outline of the book

Now introduce the chapters that follow. The presentation is done in a rather technical way instead of merely by explanations. This way, the reader with some knowledge of $\mathcal{H}_\infty$ control, model matching, IMC, etc. will recognise better if a particular chapter is of primary interest. As each chapter can be addressed in an almost self-contained way. The reading of the book does not need to be done in a purely sequential way.

In this presentation, it has to be understood that some terms and symbols are defined within the corresponding chapter. Therefore, here the authors rely on the previous knowledge the reader may have as a proof of the readiness to go directly to such chapter.

## Chapter 2: Simple Model-Matching Approach to Robust PID Design

The following observation can be made regarding the design in [81] that is the basic starting point of the approach presented in this book for solving the PID controller design problem as a particular form of 1.15:

- The tuning parameters $z$ and $T_M$ have a very similar effect on the final controller and are somehow redundant.

This means that the settings used for analytical derivation can be simplified to make the final solution dependent on a single tuning parameter. The corresponding simplified design can be found in Chapter 2. In particular, the settings below are suggested:

- $T_d = 1$. For this reference model, the problem is a sensitivity problem, see Figure 1.4. Now, $e$ is the conventional error between the reference and the actual output, and problem (1.15) is a Weighted Sensitivity Problem (WSP):

$$\min_{Q \in \mathcal{RH}_\infty} \|WS\|_\infty \tag{1.24}$$

- $W = \frac{1}{s}$. This weight is used to force integral action ($S(0) = 0$). Note that the resulting performance objective is

$$\min \left\| S \frac{1}{s} \right\|_\infty = \min \max_\omega \left| S(j\omega) \frac{1}{j\omega} \right| \tag{1.25}$$

Although (1.25) has often been used as a general performance criterion for control design, it is particularly suitable for the servo mode. This can be understood from (1.2). The relation between the reference $r$ and the error $e = r - y$ is given by $T_{er} = S = \frac{1}{1+PK}$. At low frequencies (where feedback is effective), $|L| = |PK| \gg 1$, and $S \approx P^{-1}K^{-1}(j\omega) \approx P^{-1}(j\omega)\frac{1}{k_i}j\omega$, where $k_i$ represents the integral gain of the controller (in case it is of PID type, $k_i = \frac{K_c}{T_i}$). Therefore, when $\omega \to 0$ in (1.25), $S(j\omega)\frac{1}{j\omega} \to P^{-1}(j\omega)\frac{1}{k_i}$.

- $P = K_g \frac{-\frac{h}{2}s+1}{(\tau s+1)(\frac{h}{2}s+1)}$, which results from using a first-order Padé approximation for the time delay in the FOPTD process model.

For this setup, the optimal $Q$ parameter is not proper, which leads to an improver $K$. To circumvent this problem, $Q$ is augmented using the filter $f = \frac{1}{(\lambda s+1)^2}$, where $\lambda$ is the only tuning parameter. The resulting design is very similar to IMC, and $\lambda$ is used in the same way to detune the optimal controller. If one assumes multiplicative uncertainty $\Delta$, the Small-gain theorem ensures the closed-loop stability provided that $|T| = |PQ| < 1/|\Delta| \ \forall \omega$. Therefore, increasing $\lambda$ reduces the closed-loop bandwidth and contributes to high-frequency robustness against model uncertainty. Another reason to keep $|T|$ small at high frequencies is that sensor noise is transferred to the output by $T$ [72]. In terms of $\lambda$, the final feedback controller is given by

$$K = \frac{1}{K_g \left(2\lambda + \frac{h}{2}\right)} \frac{(\frac{h}{2}s + 1)(\tau s + 1)}{s\left(\frac{\lambda^2}{2\lambda+\frac{h}{2}}s + 1\right)} \tag{1.26}$$

Note that the gain of the controller $K$ (i.e., $|K(j\omega)|$) gets reduced by increasing $\lambda$. Because $T_{ur} = KS$, large values of $\lambda$ will yield moderate levels of control activity. Taking all these considerations into account, $\lambda$ is finally fixed to get an automatic rule that directly gives the controller parameters in terms of the process model information. The choice $\lambda = h$, which results into the ISA-PID tuning rule

$$
\begin{aligned}
K_c &= \frac{0.4 T_i}{K_g h} \\
T_i &= \tau + 0.1 h \\
\frac{T_d}{N} &= 0.4 h \\
N + 1 &= 1.25 \frac{\tau}{T_i}
\end{aligned}
\tag{1.27}
$$

yields very similar results to (1.23). Both (1.23) and (1.27) are aimed at *smooth* set-point response, yielding $M_S \approx 1.42$, which is a good robustness indicator[6]. By *smooth* control we mean the slowest possible control with acceptable disturbance rejection [70]. Of course, the definition may depend on the application at hand, and the term *smooth* is sometimes used in this book merely as a synonym of *robust*.

## Chapter 3: Alternative Design for Load Disturbance Improvement

A common drawback of [81] and the design in Chapter 2 is that poor disturbance attenuation is obtained for *lag-dominant* plants[7]. This phenomenon occurs because of pole-zero cancellation between the plant $P$ and the controller $K$. Effectively, both (1.19) and (1.26) have a zero at the plant pole at $s = -1/\tau$. In order to prevent this to happen, a modification of [81] is presented in [82] which consists of specifying a different target closed-loop, taken now as:

$$
T_d = \frac{(T_M - \gamma)s + 1}{1 + T_M s}
\tag{1.29}
$$

Equivalently, the new $T_d$ corresponds to the desired sensitivity function

$$
S_d = 1 - T_d = \frac{\gamma s}{T_M s + 1}
\tag{1.30}
$$

With this change, the solution to (1.15) is

$$
K = \frac{1}{K_g(\rho + \gamma)} \frac{(1 + \tau s)(1 + \chi s)}{s(1 + \frac{z T_M + h\chi}{\rho + \gamma} s)}
\tag{1.31}
$$

where

$$
\rho = \frac{h + z}{h + T_M}(h + T_M - \gamma) \qquad \chi = h + z - \rho + T_M - \gamma
\tag{1.32}
$$

Note that when $\gamma = T_M$, the design coincides with that revised in Section 1.2.3. In particular, (1.31) and (1.32) simplify to (1.19) and (1.21). By fixing $z, T_m$ as in (1.22), $\gamma$

---

[6] Mid-frequency sensitivity to modelling errors can be captured by the peak of the sensitivity function:

$$
M_S \doteq \|S(j\omega)\|_\infty \doteq \max_\omega \left| \frac{1}{1 + L(j\omega)} \right|
\tag{1.28}
$$

$M_S$ indicates the inverse of the shortest distance from the Nyquist plot to the critical point $-1 + 0j$ [72].

[7] With respect to the FOPTD model (1.16), a process is said to be lag-dominant if $\tau/h$ is larger than about 10 (integrating plants fit into this category as an extreme example). Sometimes, however, by lag-dominant we just mean $\tau/h > 1$, including plants with relatively balanced lag/delay ratio.

turns out to be the only tuning parameter now. The role of $\gamma$ is to balance the performance between the servo and regulator modes:

- Setting $T_M = \sqrt{2}$ corresponds to the servo mode (good set-point tracking).

- Setting $T_M > \sqrt{2}$ allows improvement of the regulatory performance (better disturbance attenuation).

The above points are fully addressed in Chapter 3, where the *servo/regulation* interval for $\gamma$ is found to be:

$$\sqrt{2}h \leq \gamma \leq \frac{12.36h(\tau - \sqrt{2}h)}{h + \tau} \tag{1.33}$$

and it is assumed that $\tau \geq 1.7262h$ so that (1.33) makes sense.

## Chapter 4: Analysis of the *Smooth/Tight–Servo/Regulation* Tuning Approaches

The design in Chapter 2 depends on a single tuning parameter: $\lambda$, used to adjust the robustness/performance trade-off. Regarding the design discussed in Chapter 3, the only tuning parameter is $\gamma$, and its role is to adjust the servo/regulation trade-off. In Chapter 4, these two strategies are compared from a balanced servo/regulation point of view: first, an interval for $\lambda$ is defined based on the *smooth* and *tight* control concepts [70, 13]:

- *Smooth* control, as already commented, is the slowest possible control with acceptable disturbance rejection. In [4], $\lambda = h$ for smooth control, yielding $M_S \approx 1.42$.

- *Tight* control is the fastest possible control with acceptable robustness. In [4], $\lambda = 0.56h$ for tight control, resulting in $M_S \approx 1.75$.

Then, the trade-off value $\lambda = 0.7h$ is picked up to be used for balanced servo/regulation operation. The conclusion of Chapter 4 is that one can use the tuning $\lambda = 0.7h$ for plants such that $h/\tau \geq 0.1$. In these cases, the $\gamma$-tuning method of Chapter 3 provides no real advantage. However, for more lag-dominant plants, the $\gamma$-tuning technique allows better suppression of load disturbances with the same degree of robustness.

## Chapter 5: $\mathcal{H}_\infty$ Design with Application to PI Tuning

An improved selection of $W$ is presented in Chapter 5. According to [10], the weight $W$ in the WSP can be chosen as follows:

$$W = \frac{(\lambda s + 1)(\gamma_1 s + 1) \cdots (\gamma_k s + 1)}{s(\tau_1 s + 1) \cdots (\tau_k s + 1)} \tag{1.34}$$

where $\tau_1, \ldots, \tau_k$ are the time constants of the unstable or slow poles of $P$, $\lambda > 0$, and

$$\gamma_i \in \left[\lambda, |\tau_i|\right] \tag{1.35}$$

As before, $\lambda$ is used to adjust the robustness/performance trade-off. The $\gamma_i$ parameters permit balancing the servo/regulation performance [10]. For the sake of clarity, set $\lambda \approx 0$ and $\tau_i > 0$ (stable plant case). We have that:

- If $\gamma_i = \tau_i, i = 1, \ldots, k$, the corresponding weight is $W = \frac{1}{s}$, and the WSP is equivalent to the performance objective $\min \left\| \frac{1}{s} S \right\|_\infty$, which is suitable for the servo mode.

- If $\gamma_i = \lambda, i = 1, \ldots, k$, the resulting weight $W$ reduces $|S|$ at low frequencies to improve the disturbance rejection properties. Note that, heuristically, the choice $\gamma_i = \lambda$ can be understood (for small values of $\lambda$ and neglecting the effect of the zeros of $P$) in terms of the performance objective $\min \left\| \frac{1}{s} T_{yd} \right\|_\infty = \min \left\| \frac{1}{s} SP \right\|_\infty$. This performance objective is suitable for regulatory purposes [39]. In particular, when $\omega \to 0$, one has that $\frac{1}{j\omega} S(j\omega) P(j\omega) \to \frac{1}{k_i}$, which is coherent with the well-known fact that the integral gain of the controller gives a measure of the system's ability to reject low-frequency load disturbances[8].

A good point of the weight (1.34) is that it is also valid for unstable plants ($\tau_i < 0$ for some $i$). As it is shown in [10, 9], the use of the possibly unstable weight (1.34) avoids any notion of coprime factorization. Let us assume that $P$ is purely rational and contains at least one RHP zero, in such a way that:

$$P = \frac{n_p}{d_p} = \frac{n_p^+ n_p^-}{d_p^+ d_p^-} \tag{1.36}$$

where $n_p^+, d_p^+$ contain the unstable (or slow in the case of $d_p^+$) zeros of $n_p, d_p$, and $n_p^-, d_p^-$ contain the stable zeros of $n_p, d_p$. The weight (1.34) can be factored similarly

$$W = \frac{n_w}{d_w} = \frac{n_w}{d_p^+ d_w'} \tag{1.37}$$

Then, the optimal weighted sensitivity using the weight (1.37) is given by

$$\mathcal{N}^o = \rho \frac{q(-s)}{q(s)} \tag{1.38}$$

where $\rho$ and $q = 1 + q_1 s + \cdots + q_{\nu-1} s^{\nu-1}$ (a hurwitz polynomial) are uniquely determined by the interpolation constraints:

$$W(z_i) = \mathcal{N}^o(z_i) \qquad i = 1 \ldots \nu, \tag{1.39}$$

being $z_1 \ldots z_\nu$ ($\nu \geq 1$) the RHP zeros of $P$. The corresponding controller is:

$$K = \frac{d_p \chi}{\rho n_p^- q(-s) d_w} = \frac{d_p^- \chi}{\rho n_p^- q(-s) d_w'} \tag{1.40}$$

where $\chi$ is a polynomial satisfying

$$q(s) n_w - \rho q(-s) d_w = n_p^+ \chi \tag{1.41}$$

Based on simple models for the plant, the described procedure can be applied to PID tuning. In particular, Table 1.1 collects the tuning rules associated with first- and second-order models. In the second order model case, it is assumed that $|\tau_1| > \tau_2 > h > 0$, and that a series form PID controller is used for implementation:

$$K = K_c \left( 1 + \frac{1}{T_i s} \right) (T_d s + 1) \tag{1.42}$$

The use of the series form is convenient here because it allows a simplification of the tuning expressions [69]; in particular, we get the simple relationship $T_d = \tau_2$ for the derivative time. For implementation purposes, the derivative filter present in the real PID forms should also be designed. In order to convert the ideal PID law into the real one in the best possible way, it is advisable to follow the indications given in [44].

---

[8] For example, for a PID controller [20] show that a unit step disturbance applied at the plant input yields an integral of the error (IE) equal to $-1/k_i$, i.e.:

$$\text{IE} = \int_0^\infty e(\tau) d\tau = -\frac{T_i}{K} = \frac{-1}{k_i}$$

Indeed, the result holds for any controller including integral action. For a robust design exhibiting a non-oscillatory response, one has that $|\text{IE}| = 1/k_i \approx \text{IAE} = \int_0^\infty |e(\tau)| d\tau$.

**TABLE 1.1**
PI/D tuning rules based on $\mathcal{H}_\infty$ weighted sensitivity.

| Model | $K_c$ | $T_i$ | $T_d$ | |
|---|---|---|---|---|
| $K_g \frac{e^{-sh}}{\tau s + 1}$ | $\frac{1}{K_g} \frac{T_i}{\lambda + \gamma + h - T_i}$ | $\frac{\tau(h + \lambda + \gamma) - \lambda \gamma}{\tau + h}$ | - | $\gamma \in [\lambda, |\tau|]$ |
| $K_g \frac{e^{-sh}}{(\tau_1 s + 1)(\tau_2 s + 1)}$ | $\frac{1}{K_g} \frac{T_i}{\lambda + \gamma + h - T_i}$ | $\frac{\tau_1(h + \lambda + \gamma) - \lambda \gamma}{\tau_1 + h}$ | $T_d = \tau_2$ | $\gamma \in [\lambda, |\tau_1|]$ |

## Chapter 6: Generalized IMC Design and $\mathcal{H}_2$ Approach

The frequency domain design in Chapter 5 uses the $\mathcal{H}_\infty$ norm. However, the same kind of WSP can be posed in terms of the $\mathcal{H}_2$ norm to make the resulting design closer to conventional IMC. This is the purpose of Chapter 6, where the $\mathcal{H}_\infty$ WSP is replaced with

$$\min_{K \in \mathcal{C}} \|WS\|_2 \tag{1.43}$$

Now, the servo/regulation modes are understood in terms of input/output disturbances. With respect to Chapter 5, there are several other differences: first, the weight depends on the input type (e.g., steps, ramps, etc); second, $P$ may contain a time delay (in this case, a dead time compensator is directly obtained). In addition, the extension to plants with complex conjugate poles is addressed; for oscillating plants, it is especially important to distinguish between the servo and regulatory tasks [38, 6].

For illustration purposes, let us assume that the inputs to the system are step signals and denote by $s = -1/\tau_1, \ldots, -1/\tau_k$ the unstable/slow poles of $P$ (we restrict here to non-repeated real poles). Then, the weight in (1.43) is taken as

$$W = \frac{(\lambda s + 1)^n (\gamma_1 s + 1) \cdots (\gamma_k s + 1)}{s(\tau_1 s + 1) \cdots (\tau_k s + 1)} \tag{1.44}$$

The only difference with (1.34) is the term $(\lambda s + 1)^n$, where $n$ is used to ensure the properness of the final controller. In the case at hand, $n$ must be at least equal to the relative degree of $P$ [6]. The other parameters: $\lambda > 0$ and $\gamma_1, \ldots, \gamma_k \in [\lambda, |\tau|]$, have the same meaning as in Chapter 5. Because the WSP problem is now posed in terms of the $\mathcal{H}_2$ norm, the performance interpretation changes accordingly; for negligible values of $\lambda$:

- If $\gamma_i = |\tau|$, $|W| \approx \left|\frac{1}{s}\right|$. In this case, the WSP (1.43) minimizes the ISE with respect to a step disturbance entering at the plant output (which is equivalent to minimizing the ISE for a set-point change).

- If $\gamma_i = \lambda$, $W \approx \frac{1}{s(\tau_1 s + 1) \cdots (\tau_k + 1)}$, and the WSP (1.43) minimizes now the ISE with respect to a step disturbance passing through the conflictive poles of $P$ (input/load disturbances).

Set $P = P_a P_m$, where $P_a \in \mathcal{RH}_\infty$ is all-pass and $P_m$ is MP. By using the IMC parameterization (1.17) for $K$, a *quasi-optimal* proper solution to (1.43) is given by:

$$Q = (P_m W)^{-1} \left\{ P_a^{-1} W \right\}_\star \tag{1.45}$$

where the operator $\{\}_\star$ denotes that after a partial fraction expansion (PFE) of the operand, all terms involving the poles of $P_a^{-1}$ are omitted[9]. Equation (1.45) can be expressed as

$$Q = P_m^{-1} f \tag{1.46}$$

---

[9]Note that the operator $\{\}_\star$ is slightly different from the operator $\{\}_\star$ defined in [49, Theorem 5.2-1]. See section 6.4 for a more precise definition.

with $f = W^{-1}\left\{P_a^{-1}W\right\}_\star$. Taking $W = \frac{n_w}{d_w}$, we can alternatively write $f$ as

$$f = \frac{\chi}{n_w} = \frac{\sum_{i=0}^{\delta(d_w)-1} a_i s^i}{(\lambda s + 1)^n \prod_{i=1}^{k}(\gamma_i s + 1)} \tag{1.47}$$

where $\delta(d_w)$ denotes the degree of $d_w$ and $a_0, \ldots, a_k$ are determined from the following system of linear equations

$$T|_{s=\pi_i} = PQ|_{s=\pi_i} = P_a f|_{s=\pi_i} = 1 \qquad i = 1 \ldots \delta(d_w) \tag{1.48}$$

being $\pi_i, i = 1, \ldots, \delta(d_w)$ the poles of $W$. As long as the $a_i$ coefficients satisfy (1.48), the filter time constants $\lambda$ and $\gamma_i$ can be selected freely without any concern for nominal stability or asymptotic tracking. By using Lagrange-type interpolation theory [49], it is possible to develop an expression for (1.47) explicitly:

$$f = \frac{1}{n_w} \sum_{j=1}^{\delta(d_w)} (P_a^{-1} n_w)|_{s=\pi_j} \prod_{\substack{i=1\\i\neq j}}^{\delta(d_w)} \frac{s - \pi_i}{\pi_j - \pi_i} \tag{1.49}$$

The filter (1.49), or (1.47), represents a generalization of the conventional filter (1.7) used within IMC. In the case at hand, the $\gamma_i$ parameters are used to balance the regulatory performance between step-like input/output disturbances.

## Chapter 7: PID Design as a Weighted Sensitivity Problem

The key point in the approach presented so far in the preceding chapters is the important role of the weight in the WSP. Chapter 6 presented a generic way of constructing weighting filters within the IMC framework. In fact, the proposed general filter takes the form:

$$W = \frac{(\lambda s + 1)^n}{d_d} \prod_{i=1}^{k} \frac{\gamma_i s + 1}{\tau_i s + 1} \tag{1.50}$$

with

$$n = \max\left\{1, \delta(d_d) + \delta(P) - 1\right\} \tag{1.51}$$

where $\delta(d_d), \delta(P)$ denote the degree of $d_d$ and the relative degree of $P$, respectively. For the common case of step disturbances ($d_d = s$), (1.51) simplifies to $n = \max\left\{1, \delta(P)\right\}$. Finally, $\lambda$ and $\gamma_1, \ldots, \gamma_k$ in (1.50) are tuning parameters verifying that

$$\lambda > 0 \quad , \quad \gamma_i \in [\lambda, |\tau_i|] \tag{1.52}$$

The use of such weight within IMC allowed for a generalised analytical solution in terms of generic IMC filters. However, we are interested in PI/PID controllers. A first step towards this direction was made in Chapter 5, where it is exemplified how, for a FOPTD model of the process given by $P_o = K_g \frac{e^{-sh}}{\tau s + 1}$, where $K_g, h, \tau$ are, respectively, the gain, the (apparent) delay, and the time constant—negative in the unstable case—, and by selecting the weight as

$$W = \frac{(\lambda s + 1)(\gamma s + 1)}{s(\tau s + 1)} \tag{1.53}$$

where $\lambda > 0, \gamma \in [\lambda, |\tau|]$. In this case, the feedback controller can be cast into the PI structure:

$$K = K_c\left(1 + \frac{1}{T_i s}\right) \tag{1.54}$$

In this chapter, this case is extended further, and the generic case of a second-order plus time delay process model is considered:

$$P = \frac{ke^{-\theta s}}{(\tau_1 s + 1)(\tau_2 s + 1)} \tag{1.55}$$

in conjunction with the following weight:

$$W = \frac{(\lambda s + 1)(\gamma_1 s + 1)(\gamma_2 s + 1)}{s(\tau_1 s + 1)(\tau_2 s + 1)} \tag{1.56}$$

where

$$\lambda > 0, \gamma_i \in [\lambda, |\tau_i|], i = 1, 2. \tag{1.57}$$

The problem is solved analytically and the general solution for the $\mathcal{H}_\infty$ PID controller is provided. This generic solution serves to generate all the corresponding particular cases that originates from process models that can be derived from (1.55). The particularisation from the SOPTD model to the more simple model (say, FOPTD, IPTD, etc) is applied also to the generic solution in order to generate the corresponding controller. All special cases are presented and tuning expressions generated.

It is worth highlighting here that all these sort of tuning relations are expressions that depend on $\gamma$ and $\lambda$ that remain the final tuning parameters that will determine the servo/regulation and robustness/performance trade-offs. The guidelines for appropriate selection of $\gamma$ and $\lambda$ will be the focus of the next chapter.

## Chapter 8: PID Tuning Guidelines for Balanced Operation

A side effect of the general PID tuning expressions in Chapter 7 was the change of dimensionality with respect to the tuning of the PID controller. From the PID definition, the tuning problem is a three-dimensional search problem. With the tuning provided in Chapter 7 the problem is a two-dimensional one. In addition to this dimensionality reduction, there is an additional advantage of the provided PID tuning with respect to the meaning of the parameters. Now, with $\gamma$ and $\lambda$, the parameters are related to the servo/regulation and robustness issues. How to balance them is the subject of this chapter. In that respect, designs are first of all put at the same level for fair comparison. Comparable designs are established with the help of fixing the robustness level. For each robustness level, in terms of $M_S$, we can define

$$\Lambda\Gamma_k \doteq \left\{(\lambda, \gamma); M_S = k, \lambda > 0, \gamma \in [\lambda, \tau]\right\} \tag{1.58}$$

The considered problem is how to select $\gamma$ to yield a good balance between servo and regulatory performance. For this purpose, we consider the minimization of two alternative performance indices:

$$J_{\max} = \max(\Delta_s, \Delta_r) \tag{1.59}$$
$$J_{\text{avg}} = 0.5(\Delta_s + \Delta_r) \tag{1.60}$$

where

$$\Delta_s = \frac{\text{IAE}_s}{\text{IAE}_s^o}, \quad \Delta_r = \frac{\text{IAE}_r}{\text{IAE}_r^o} \tag{1.61}$$

Then, for each robustness level ($M_S = k$), we can consider the following optimization problem

$$\min_{(\lambda,\gamma)\in\Lambda\Gamma_k} J \qquad\qquad (1.62)$$

In this chapter, we provide tuning guidelines for the $\lambda, \gamma$ parameters using the performance indices $J_{\max}, J_{\mathrm{avg}}$ and the optimization problem (1.62). The overall objective is to achieve a *balanced* closed-loop, that is, a good balance between servo and regulatory performance, on the one hand, and between robustness and performance, on the other hand. For each one of the particular cases provided in the previous chapter, the optimisation problem is solved and the ranges for $\lambda, \gamma$ analyzed.

For auto-tuning purposes, a robustness level of $M_S = 1.6$ is selected as a good trade-off between robustness and performance for the considered indices. Then, tuning guidelines for $\lambda, \gamma$ in terms of the process model parameters are provided. The specific cases considered in Chapter 7 are revisited and, for each one, the optimal solutions with respect to $J_{\max}, J_{\mathrm{avg}}$ are presented.

# Part I

# Model-Matching Approach to Robust PID Design

# 2

## Simple Model-Matching Approach to Robust PID Control

This chapter addresses PID tuning for robust set-point response from a min-max model-matching formulation. Within the considered context, several setups result in a PID controller. Here, we investigate the simplest one, leading to a PID controller solely dependent on a single design parameter. Attending to common performance/robustness indicators, the free parameter is finally fixed to provide an automatic tuning in terms of the model information. Simulation examples are given to evaluate the proposed settings.

### 2.1 Problem statement

In this section, the control framework and the model-matching problem (MMP) on which the controller derivation is based are introduced. The latter obeys to a min-max optimization problem that captures the performance objective.

#### 2.1.1 The control framework

The customary unity feedback controller is depicted in Figure 2.1. Closed-loop performance and robustness are typically evaluated in terms of the sensitivity $S$ and the complementary sensitivity $T$ transfer functions [72], respectively:

$$S \doteq \frac{1}{1+L} \tag{2.1}$$

$$T \doteq 1 - S = \frac{L}{1+L} \tag{2.2}$$

where $L \doteq PK(s)$ is the loop transfer function. The model of the plant is given by:

$$P = K_g \frac{e^{-sh}}{\tau s + 1} \tag{2.3}$$

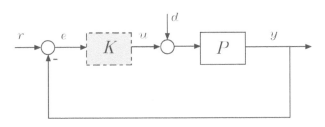

**FIGURE 2.1**
Conventional feedback configuration.

For design purposes, it is convenient to approximate the delay term in (2.3) so as to achieve a purely rational process model. By using the first-order Pade expansion

$$e^{-sh} \approx \frac{-(h/2)s + 1}{(h/2)s + 1} \tag{2.4}$$

model (2.3) can be approximated as follows

$$P \approx K_g \frac{-\frac{h}{2}s + 1}{(\tau s + 1)(\frac{h}{2}s + 1)} \tag{2.5}$$

Regarding the control law, the following ISA PID form [20] is chosen:

$$u = K_c \left( 1 + \frac{1}{sT_i} + \frac{sT_d}{1 + sT_d/N} \right) e \tag{2.6}$$

where $e = r - y$, being $r$, $y$ and $u$ the Laplace transforms of the reference, process output and control signal, respectively. $K_p$ is the PID gain, whereas $T_i$ and $T_d$ are its integral and derivative time constants. Finally, $N$ is the ratio between $T_d$ and the time constant of an additional pole introduced to assure the properness of the controller. This way, the following transfer function for the controller $K$ is assumed:

$$K = K_c \frac{1 + s(T_i + \frac{T_d}{N}) + s^2 T_i \frac{T_d}{N}(N + 1)}{sT_i(1 + s\frac{T_d}{N})} \tag{2.7}$$

### 2.1.2   The model-matching problem

The controller design will be based on a desired input-output response. Mathematically, the following min-max optimization problem is posed to capture the performance objective:

$$\min_{K \in \mathcal{C}} \left\| W(T_d - T) \right\|_\infty \tag{2.8}$$

where $T_d$ is a desired reference model for the closed-loop system response, $W$ is a weighting function and $T$ is the complementary sensitivity function, which corresponds to the transfer function from the input to the output. In Section 2.2 the control problem (2.8) will be solved for a suitable particular case yielding a regulator $K$ of the form (2.7). The Youla parametrization [49] for stable plants will be used to simplify the search of the optimal stabilizing controller in (2.8). According to Figure 2.1, this result states that any *internally stabilizing* controller $K$ can be expressed as

$$K = \frac{Q}{1 - PQ} \tag{2.9}$$

where $Q$ is any stable transfer function. The role of $Q$ is better understood within the IMC configuration [49] depicted in Figure 2.2. In the context of the IMC structure, $Q$ is the parameter to be designed. The main advantage of this approach comes from the fact the all the closed-loop feedback relations become *affine* in the $Q$ parameter. For instance, the closed-loop mapping, $H(P, K)$, in 1.1 simplifies to:

$$\begin{pmatrix} y \\ u \end{pmatrix} = \begin{pmatrix} PQ & P(1 - PQ) \\ Q & -PQ \end{pmatrix} \begin{pmatrix} r \\ d \end{pmatrix} \tag{2.10}$$

In particular, one has that:

$$S = 1 - PQ \tag{2.11}$$

and

$$T = PQ \tag{2.12}$$

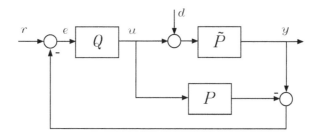

**FIGURE 2.2**

IMC configuration. Here, $\tilde{P}$ represents the real (uncertain) plant, whereas $P$ denotes the available model. In the nominal scenario, the model is assumed to be perfect, i.e., $\tilde{P} = P$.

With all these considerations in mind, the constrained problem (2.8) can be posed in terms of $Q$ as follows:

$$\min_{Q \in \mathcal{RH}_\infty} \left\| W \left(T_d - PQ\right)\right\|_\infty \tag{2.13}$$

Once the problem above has been solved, the equivalent unity feedback controller is obtained from (2.9).

### 2.1.3 The model-matching problem within $\mathcal{H}_\infty$ control

As stated, the (SISO) MMP [26] consists of:

$$\min_{Q \in \mathcal{RH}_\infty} \left\| T_1 - T_2 Q \right\|_\infty \tag{2.14}$$

where $T_1, T_2 \in \mathcal{RH}_\infty$. Let $\{z_1, z_2, \ldots, z_\nu\}$ be the distinct right half-plane (RHP) zeros of $T_2$. Then,

**Lemma 2.1.1.** *The optimal matching error $\mathcal{E}^o = T_1 - T_2 Q$ in (2.14) is an all-pass function [28, 26, 81], more precisely:*

$$\mathcal{E}^o(s) = \begin{cases} \rho \dfrac{q(-s)}{q(s)} & \text{if } \nu \geq 1 \\ 0 & \text{if } \nu = 0 \end{cases} \tag{2.15}$$

*where $q(s) = 1 + q_1 s + \cdots + q_{\nu-1} s^{\nu-1}$ is a strictly Hurwitz polynomial. Furthermore, the constants $\rho$ and $\{q_i\}_{i=1}^{\nu-1}$ are real and are uniquely determined by the interpolation constraints*

$$\mathcal{E}(z_i) = T_1(z_i) \quad i = 1 \ldots \nu \tag{2.16}$$

**Remark 2.1.1.** *For $\nu = 1$ or $\nu = 2$, Lemma 2.1.1 can be directly used to obtain explicit formulae for the solution of (2.14) [90]. However, because the interpolation conditions (2.16) constitute a nonlinear system, for $\nu \geq 3$ it should be better to find $\mathcal{E}^o$ by using a more systematic procedure like the Nevanlinna-Pick's algorithm [26].*

The importance of the MMP relies on the fact that any generalized control problem can be expressed as a MMP[1]. This is achieved by means of the celebrated Youla-Kucera

---

[1]In general, in the multivariable case, the MMP is expressed as $\min_{Q \in \mathcal{RH}_\infty} \left\| T_1 - T_2 Q T_3 \right\|_\infty$, where $Q, T_1, T_2, T_3 \in \mathcal{RH}_\infty$ are matrix transfer functions.

parameterization [89, 80, 72], which states that any $K \in \mathcal{C}$ can be parameterized as

$$K = \frac{Y + MQ}{X - NQ} \tag{2.17}$$

being $Q \in \mathcal{RH}_\infty$ a free parameter, and $X, Y, M, N \in \mathcal{RH}_\infty$ a coprime factorization of $P$, implying that $P = NM^{-1}$ and $XM + YN = 1$ [80, 72, 40]. The above result allows the expression any closed-loop transfer function affinely in $Q$ in the form of a MMP. For example, a basic problem in $\mathcal{H}_\infty$ (of main importance in this book) is the WSP [72, Section 2.8.2]:

$$\min_{K \in \mathcal{C}} \|WS\|_\infty \tag{2.18}$$

where $W$ is a performance weight in charge of shaping $S$ conveniently. With respect to Figure 1.2, the WSP corresponds to the generalized plant

$$G = \left[ \begin{array}{c|c} W & -WP \\ \hline 1 & -P \end{array} \right] \tag{2.19}$$

Using (2.17), the sensitivity function can be parameterized as $S = M(X - NQ)$, and the WSP (2.18) can then be cast into the MMP form by selecting $T_1 = WMX, T_2 = WMN$, where $T_1, T_2 \in \mathcal{RH}_\infty$ as long as $W \in \mathcal{RH}_\infty$.

## 2.2   Analytical solution

We are now concerned with finding a simple solution to problem (2.13), which aims at minimizing the functional

$$\mathcal{E} = \|W(T_d - T)\|_\infty \tag{2.20}$$

Several methods could be followed in order to solve this $\mathcal{H}_\infty$ general problem. See, for example, [28, 85]. However, our interest focuses on simple instances of the problem (2.8) leading to a controller of the form (2.7). Following this rationale, the above min-max problem was solved in [81] for the following particular setup:

$$W = \frac{1 + zs}{s} \qquad T_d = \frac{1}{1 + T_M s}$$

Additionally, a first-order Taylor approximation for the delay in (2.3) was taken into account. The resulting controller depends on two tuning parameters: $z, T_M$. The role of $T_M$ is clear: it specifies the desired speed of response, whereas $z$ allows for adjustment of the robustness of the control system.

Two different setups are presented here for solving this initial problem. The first section presents the setup based on the initial work of [81], whereas the next one slightly changes the setup in order to present a step for stability analisys in subsequent sections as well as formulation improvements in the next chapters.

### 2.2.1   Initial formulation for set-point response

Let us consider the following problem

$$\min_{K \in \mathcal{C}} \|W(T_d - T)\|_\infty \tag{2.21}$$

where $T_d$ represents the desired complementary sensitivity (the input-to-output response: $T_{yr}$) and $W$ is a frequency weight (see Figure 2.3).

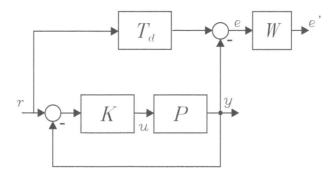

**FIGURE 2.3**
Diagram for problem (2.21). Here, $e$ denotes the error between the desired and actual outputs.

The idea in [81] is to use simple settings in order to obtain a PID controller:

- $T_d = \frac{1}{T_M s + 1}$, where the $T_M$ parameter specifies the desired *speed of response*.

- $W = \frac{zs+1}{s}$. The integrator forces integral action by requiring perfect matching between $T$ and $T_d$ at zero frequency. The $z$ parameter is used to adjust the robustness margins: the larger the value of $z$, the more robust the resulting system.

In addition, many processes have rather simple dynamics and are often modeled using low-order models of the form

$$P = K_g \frac{e^{-sh}}{\tau s + 1} \tag{2.22}$$

called first-order plus time delay or just FOPTD systems, where $K_g, h, \tau$ are, respectively, the gain, the (apparent) delay, and the time constant of the process. These models can be obtained easily through open-loop and closed-loop step response tests [66]. Alternatively, one can start from an accurate description of the process and apply some model reduction technique. In this regard, Skogestad's *half-rule* [69] provides a simple analytic approach. Supported by these considerations, a stable FOPTD model is used in [81]. However, for derivation purposes, the time delay in (2.22) is approximated using a first-order Taylor expansion:

$$e^{-sh} \approx 1 - sh$$

that generates the following approximated plant model:

$$P = K_g \frac{-sh + 1}{\tau s + 1} \tag{2.23}$$

Note that the problem at hand could now be posed in terms of the general control setup (see Figure 2.4) to be solved numerically for particular values of the tuning parameters. For such a simple problem, however, the solution can be obtained analytically. First, note that for stable plants we can take $X = 1, M = 1, N = P, Y = 0$ in (2.17), and express all stabilizing controllers as

$$K = \frac{Q}{1 - PQ} \tag{2.24}$$

Then, in terms of $Q$, $T = PQ$, and (2.21) is equivalent to

$$\min_{Q \in \mathcal{RH}_\infty} \left\| W(T_d - PQ) \right\|_\infty \tag{2.25}$$

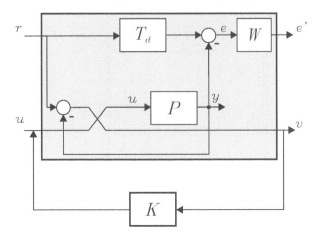

**FIGURE 2.4**
Problem (2.21) rearranged in the general control form.

which is a MMP (2.14) with $T_1 = WT_d, T_2 = WP$. The solution to (2.25) follows immediately from Lemma 1.2.1 (with $\nu = 1$) and, after straightforward algebra, the resulting feedback compensator obtained using (2.24) turns out to be a PID controller:

$$K = \frac{1}{K_g(\rho + T_M)} \frac{(1 + \tau s)(1 + \chi s)}{s(1 + \frac{zT_M + h\chi}{\rho + T_M}s)} \tag{2.26}$$

where

$$\rho = \frac{h + z}{h + T_M}h \qquad \chi = h + z - \rho \tag{2.27}$$

The tuning parameters $T_m, z$ can be fixed for auto-tuning purposes as in [81]:

$$T_M = 2h \qquad z = \sqrt{2}h \tag{2.28}$$

With these values for $z, T_M$, which are chosen according to robustness considerations, the tuning rule (in ISA-PID form) can be expressed as

$$
\begin{aligned}
K_c &= \frac{T_i}{K_g h 2.65} \\
T_i &= \tau + 0.03h \\
\frac{T_d}{N} &= 1.72h \\
N + 1 &= \frac{\tau}{T_i}
\end{aligned}
\tag{2.29}
$$

Vilanova's tuning rule [81] has been found to work well in practice, exhibiting a robust behaviour in the face of process abnormalities including model mismatch, valve stiction, and sensor noise [14].

## 2.2.2   Alternative formulation

In what follows, a slightly different setup is suggested, resulting in a single-parameter PID control law which provides very similar performance. In this case, we improve the delay approximation but use a more simple formulation for the weighting function.

The suggested settings for minimizing (2.20) are:

$$P = K_g \frac{-\frac{h}{2}s + 1}{(\frac{h}{2}s + 1)(\tau s + 1)} \qquad T_d = 1 \qquad W = \frac{1}{s} \tag{2.30}$$

The weight $W = \frac{1}{s}$ is the simplest one ensuring integral action in the design, whereas the selected reference model $T_d = 1$ specifies the ideal input-to-output relation. Needless to say, this is not achievable in practice: if a very quick response is desired, this would be normally at the expense of a large overshoot in the output transient and poor robustness margins. As it will be seen, controlling the overshoot will be an easy task once the optimum controller has been derived. Substituting the expressions for $W$ and $T_d$ into (2.13) we finally arrive at

$$\min_{Q \in \mathcal{RH}_\infty} \left\| \frac{1}{s}(1 - PQ) \right\|_\infty \tag{2.31}$$

Note that problem (2.31) is indeed a sensitivity one because $1 - PQ = S$. In addition, (2.31) corresponds to a MMP with $T_1 = 1/s, T_2 = P/s$. Because $T_2$ (equivalently $P$) has only one RHP zero ($\nu = 1$) at $s = h/2$, Lemma 2.1.1 implies that the optimal $\mathcal{E}$ in (2.20) is[2]

$$\mathcal{E}^o = \rho \frac{q(-s)}{q(s)} = \rho \tag{2.32}$$

where the constant $\rho$ is determined from the interpolation constraint

$$\mathcal{E}\left(\frac{2}{h}\right) = W\left(\frac{2}{h}\right) T_d\left(\frac{2}{h}\right) = W\left(\frac{2}{h}\right) = \frac{h}{2} \tag{2.33}$$

Consequently, from (2.33), $\rho = \frac{h}{2}$. This means that the optimal IMC controller $Q$ is such that

$$\mathcal{E}^o = \rho = \frac{h}{2} = W(T_d - PQ) = \frac{1}{s}(1 - PQ) \tag{2.34}$$

By isolating $Q$ from (2.34) we arrive at

$$Q = \left(-\frac{h}{2}s + 1\right) P^{-1} = \frac{1}{K_g}\left(\frac{h}{2}s + 1\right)(\tau s + 1) \tag{2.35}$$

From (2.35), we see that the optimal $Q$ solving the minimization problem is not proper. In order to yield a realizable compensator, it is necessary to augment it with a filter of appropriate order:

$$f(s) = \frac{1}{(\lambda s + 1)^2} \tag{2.36}$$

therefore

$$Q = \frac{1}{K_g}\left(\frac{h}{2}s + 1\right)(\tau s + 1)f(s) = \frac{1}{K_g}\frac{(\frac{h}{2}s + 1)(\tau s + 1)}{(\lambda s + 1)^2} \tag{2.37}$$

By making $\lambda \to 0$, the optimal behaviour is recovered. The equivalent unity feedback controller $K$ is given by (2.9)

$$K = \frac{1}{K_g}\frac{(\frac{h}{2}s + 1)(\tau s + 1)}{s(\lambda^2 s + 2\lambda + \frac{h}{2})} \tag{2.38}$$

---

[2]For the case of a single RHP zero in $T_2$, the MMP (2.14) can also be solved by direct application of the maximum modulus principle of complex variable [24, 72, 7].

and can be cast into the commercial form (2.7) according to the following tuning rule:

$$K_c = \frac{\chi(\lambda)}{K_g(4\lambda + h)} \tag{2.39}$$

$$T_i = \frac{\chi(\lambda)}{2}$$

$$\frac{T_d}{N} = \frac{2\lambda^2}{4\lambda + h}$$

$$N + 1 = \frac{\tau h(4\lambda + h)}{2\lambda^2 \chi(\lambda)}$$

where

$$\chi(\lambda) = \frac{4\lambda(2\tau + h) + (\tau + h)^2 - \tau^2 - 4\lambda^2}{4\lambda + h} \tag{2.40}$$

**Remark 2.2.1.** *Although the focus in this chapter is on the FOPTD model, the above analytical procedure is readily applicable to other low-order stable[3] models:*

- *Second-Order Processes with Time Delay (SOPTD):*
  $K_g \frac{e^{-hs}}{(\tau_1 s + 1)(\tau_2 s + 1)} \approx K_g \frac{-hs + 1}{(\tau_1 s + 1)(\tau_2 s + 1)}.$

- *Second-Order Processes with Inverse Response (SOPIR):* $K_g \frac{-\alpha s + 1}{(\tau_1 s + 1)(\tau_2 s + 1)}.$ *The latter case was considered in [7].*

## 2.3   Stability analysis

This section addresses how the $\lambda$ parameter influences both the nominal and the robust stability of the proposed controller. The main objective is to prepare the groundwork for assisting in the selection of the tuning parameter based on robustness considerations. This task is finally accomplished in Section 2.4, where a *free-of-$\lambda$* tuning rule is proposed.

### 2.3.1   Nominal stability

Since we have considered the approximation (2.5) for the adopted FOPTD model, the basic requirement of nominal stability is not guaranteed for a FOPTD plant even when all its parameters are perfectly known. The nominal stability issue is dealt with here by means of the Dual Locus technique along the lines of [94]. This technique, based upon the Argument Principle [24], can be regarded as a *modified* version of the well-known Nyquist criterion [72]. Taking the controller from (2.38) together with the model (2.3) results in the following characteristic equation:

$$1 + L = 1 + \frac{\frac{h}{2}s + 1}{\lambda^2 s^2 + (2\lambda + \frac{h}{2})s} e^{-sh} = 0 \tag{2.41}$$

which can be rewritten in the form

$$L_1 - L_2 = 0 \tag{2.42}$$

---

[3]The design, as it has been presented here, is not applicable to unstable plants since it would result in an unstable pole/zero cancellation between the plant $P$ and the controller $K$.

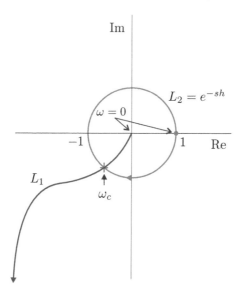

**FIGURE 2.5**
Dual locus diagram.

by making the following assignments:

$$L_1 = -\frac{\lambda^2 s^2 + (2\lambda + \frac{h}{2})s}{\frac{h}{2}s + 1} \qquad L_2 = e^{-sh}$$

The dual locus diagram technique is now applied: briefly stated, the closed-loop system is stable if the locus of $L_1$ reaches the intersection point earlier than $L_2$. The loci of $L_1$ and $L_2$ have been displayed in Figure 2.5 for positive frequencies of the imaginary axis.

The intersection frequency can be determined by solving the equation:

$$\left| -\frac{\lambda^2 s^2 + (2\lambda + \frac{h}{2})s}{\frac{h}{2}s + 1} \right|_{s=j\omega} = 1 \tag{2.43}$$

from which the positive frequency of interest can be seen to be:

$$\omega_c = \frac{\sqrt{-4 - 2\mu + \sqrt{(-4 - 2\mu)^2 + 4}}}{\sqrt{2}\lambda} \tag{2.44}$$

where $\mu = \frac{h}{\lambda}$. The phase angles of $L_1$ and $L_2$ at $\omega_c$ are, respectively:

$$\phi_1 = \arctan -\frac{2 + \frac{1}{2}\mu}{\lambda\omega_c} - \arctan \frac{1}{2}\mu\lambda\omega_c \tag{2.45}$$

and

$$\phi_2 = -h\omega_c \tag{2.46}$$

The stability condition is satisfied only when the phase angle of $L_1$ is larger (in absolute value) than that of $L_2$ at $\omega_c$, i.e, if $\phi_1 - \phi_2 < 0$. From the fact that $\mu = \frac{h}{\lambda}$ and (2.44) it can be seen that the function $\phi_1 - \phi_2$ is ultimately only a function of $\mu$.

The function $\phi_1 - \phi_2$ is plotted against the $\mu$ parameter in Figure 2.6. It can be concluded that the resulting closed loop system is stable provided that $\lambda = \frac{h}{\mu}$ is chosen such that

$$\lambda > \frac{1}{13.5135}h \approx 0.074h \tag{2.47}$$

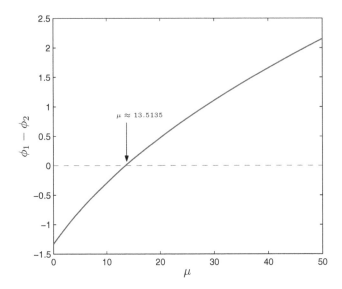

**FIGURE 2.6**
$\phi_1 - \phi_2$ vs $\mu = \frac{h}{\lambda}$.

### 2.3.2 Robust stability

To account for model uncertainty we will assume that the dynamic behaviour of a plant is described not only by a single LTI model but by a whole family, usually referred to as the *uncertainty set*. More precisely, we will consider the possible plants at hand belonging to the following set:

$$\mathcal{F} = \left\{ \tilde{P} = P(1 + \Delta_m) \right\} \tag{2.48}$$

where $\Delta_m$ is the relative (multiplicative) model error

$$\Delta_m \doteq \frac{\tilde{P} - P}{P} \tag{2.49}$$

satisfying $|\Delta_m(j\omega)| \leq |W_m(j\omega)|$. $W_m(s)$ is a frequency weight bounding the model error (plant/model mismatch). It is well-known [72, 49] that a controller $K$ that stabilizes the nominal plant $P$, also stabilizes all the plants in (2.48) provided that

$$\|W_m T\|_\infty < 1 \tag{2.50}$$

Condition (2.50) evaluates robust stability in terms of the nominal complementary sensitivity function $T$. We need to compute the relative model error between the model (2.3) and the real plant, which is considered to be

$$\tilde{P} = K_g(1 + r_k)\frac{e^{-h(1+r_h)s}}{(1 + r_\tau)\tau s + 1} \tag{2.51}$$

for $r_k, r_\tau, r_h$ in the interval $(-1, +1)$. It can be seen that if we denote by $\delta_k, \delta_\tau, \delta_h$ the maximum (positive) values of $r_k, r_\tau, r_h$, respectively, then the worst-case relative error $\Delta_m$, corresponding to the most difficult plant to stabilize, is given by:

$$\Delta_m^* = (1 + \delta_k)\frac{\tau s + 1}{(1 - \delta_\tau)\tau s + 1}e^{-sh\delta_h} - 1 \tag{2.52}$$

From (2.5) and (2.37), the nominal complementary sensitivity function is

$$T = PQ = \frac{-\frac{h}{2}s + 1}{(\lambda s + 1)^2} \tag{2.53}$$

In our case, the robust stability condition (2.50) holds if and only if

$$|T(j\omega)| < \frac{1}{|\Delta_m^*(j\omega)|} , \forall \omega \tag{2.54}$$

From the nominal stability analysis we know that by choosing $\lambda > 0.074$ the closed-loop is stable for a perfectly known FOPTD system. The necessary minimum $\lambda$—thus, providing the fastest response—yielding robust stability can be determined graphically by plotting the magnitudes of (2.53) and (2.52) for a given parametric uncertainty pattern, increasing $\lambda$ until (2.54) is satisfied. This method could be followed by the control system designer in order to conveniently adjust the robustness/performance trade-off. In order to make this procedure completely automatic, the following section proposes a way to fix $\lambda$, providing thus an auto-tuning of the proposed controller.

## 2.4 Automatic PID tuning derivation

This section is aimed at conveniently fixing the value of $\lambda$ in the tuning relations (2.39), giving rise to a tuning rule solely dependent on the model. So far, the particular MMP (2.31) has been solved. Its solution (2.39) has been found to depend on the FOPTD model in addition to an extra tuning parameter: $\lambda$. The lower the value of $\lambda$, the lower the value of the functional (2.20). However, an excessively low value for $\lambda$ providing very fast responses is not desirable since it is bound to produce large overshoots in the step response. This is not taken into account by the adopted performance criterion. Besides, from the stability analysis of the previous section, in order to make the closed loop robust, $\lambda$ has to provide the necessary detuning and cannot be so small in practice. In accordance with this, we summarize below the requirements to be met:

- *Performance:* A sufficiently fast, *free-of-overshoot* nominal set-point response. This performance specification obeys the fact that in many processes such as chemical or mechanical systems an excessive overshoot is not acceptable. Consequently, $\lambda$ has to produce a small value for the functional (2.20) while ensuring smooth set-point response.

- *Robustness:* As the controller is obtained from the model, it has to be chosen in such a way that the closed-loop is not too sensitive to variations in process dynamics. Making direct use of the robust stability condition (2.54) is not easy and would be restricted to parametric uncertainty. Instead, a more general and simpler robustness measure will be used. In spite of this, condition (2.54) will be used at a later stage to assess robustness in the face of parametric uncertainty. Sensitivity to modelling errors can alternatively be captured by the peak of the sensitivity function:

$$M_S \doteq \|S(j\omega)\|_\infty \doteq \max_\omega \left| \frac{1}{1 + L(j\omega)} \right| \tag{2.55}$$

which indicates the inverse of the shortest distance from the Nyquist plot to the critical point. Having $M_S < 2$ is a traditional robustness indicator [72].

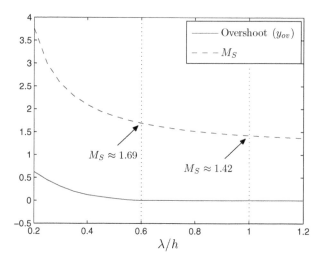

**FIGURE 2.7**
Output overshoot ($y_{ov}$) and sensitivity peak ($M_S$) against $\lambda/h$.

It is evident that both the overshoot and the sensitivity peak will depend on the loop function $L$. On the other hand, as

$$T = \frac{L}{1 + L} = \frac{-(h/2)s + 1}{(\lambda s + 1)^2}$$

is a function of just $h$ and $\lambda$, $L$ depends only on $h, \lambda$ as well. Consequently, if we define $q_1 = "overshoot"$, $q_2 = "sensitivity \ peak"$ it is clear that there exist functional relations $f_1, f_2$ such that $q_i = f_i(h, \lambda)$, $i = 1, 2$. In these functional relations we have two variables and only one independent unit (time). By applying the Buckingham pi theorem from dimensional analysis, consult for instance [77, 21], it is possible to describe the same relationships by using only one dimensionless parameter. In particular, relations $q_i = f_i(h, \lambda)$ can be expressed more compactly as $\pi_i = \phi_i(\frac{\lambda}{h})$, where $\pi_i$ contains the quantity of interest $q_i$, proving that both the overshoot and the sensitivity peak depend only on $\frac{\lambda}{h}$. This dependence can be seen in Figure 2.7, from which the zero overshoot requirement is met for $\lambda > 0.6h$.

However, at $\frac{\lambda}{h} = 0.6$ the sensitivity peak curve slope is still significant. For the sake of an improvement in robustness, some extra nominal performance in terms of closed-loop bandwidth is sacrificed by choosing $\lambda = h$, a point on which $M_S \approx 1.42$, and the sensitivity curve has a slope of almost zero. This indicates that it is not worth slowing down the nominal response further. With the choice $\lambda = h$, the tuning rule (2.39) becomes that of Table 2.1.

**TABLE 2.1**
Proposed ISA PID tuning.

| $\mathbf{K_c}$ | $\mathbf{T_i}$ | $\mathbf{T_d/N}$ | $\mathbf{N+1}$ |
|---|---|---|---|
| $\frac{0.4T_i}{K_g h}$ | $\tau + 0.1h$ | $0.4h$ | $1.25\frac{\tau}{T_i}$ |

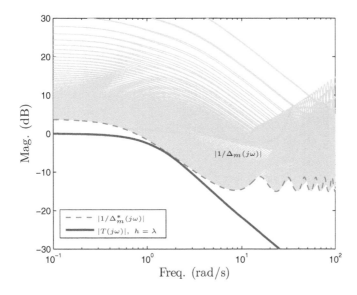

**FIGURE 2.8**
Robust stability condition for the nominal plant $\frac{e^{-0.63s}}{s+1}$, assuming 65% of parametric uncertainty in $K_g, \tau, h$.

Condition (2.54) can be used now to give an idea of the achieved robustness with $\lambda = h$ in terms of parametric uncertainty. For $\lambda = h$, (2.53) becomes

$$T = \frac{-\frac{h}{2}s + 1}{(hs + 1)^2} \tag{2.56}$$

Let us consider the worst-case uncertainty $\Delta_m^*$ in (2.52) with $\delta_k = \delta_\tau = \delta_h = \delta = 0.65$—i.e., assume 65% of simultaneous parametric uncertainty—and $T$ in (2.56) and define the variable $q_3=$*frequency distance between bode plots of $T$ and $1/\Delta_m^*$*. It is clear that $q_3 = f(h, \tau)$. By invoking again the Buckingham pi theorem, the same functional relationship can be expressed in the more compact form $\pi_3 = \phi(h/\tau)$, where $\pi_3$ contains the quantity of interest $q_3$. One can try out different values for $h/\tau$ until the magnitude bode plots of $1/\Delta_m^*$ and $T$ almost intersect. This has been found to happen for $h/\tau = 0.63$, see Figure 2.8. Consequently, it can be claimed that 65% of parametric uncertainty is allowed for any FOPTD system for which $h/\tau = 0.63$.

The described procedure can be repeated for different values of $\delta$. This experiment yields the bounds shown in Table 2.2.

It is worth noting that the proposed tuning rule is robust for lead-dominant systems, tolerating almost 100% of uncertainty in the plant parameters. Proceeding likewise, the worst-case overshoots for different values of $\delta$ have been plotted in Figure 2.9.

**TABLE 2.2**
Permissible simultaneous parametric uncertainty in $K_g, \tau, h$.

| $h/\tau$ | 0.1 | 0.25 | 0.5 | 0.63 | 1 | 5 | 10 |
|---|---|---|---|---|---|---|---|
| $\delta \times 100$ | 38% | 45% | 59% | 65% | 84% | 97% | 98% |

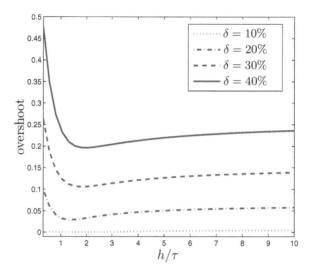

**FIGURE 2.9**
Worst-case overshoots for different levels of parametric uncertainty.

### 2.4.1   Control effort constraints

As it will be seen in Section 2.5, moderate control usage is required by the proposed controller in Table 2.1. However, as the control effort is important in practical applications, we provide here quantitative guidelines for selecting $\lambda$ according to saturation limits and slew rate constraints.

From (2.10) and (2.37), the control signal associated with a set-point change is given by

$$u(t) = \mathcal{L}^{-1}\left\{Q\frac{1}{s}\right\} = \mathcal{L}^{-1}\left\{\frac{1}{K_g}\frac{(\tau s+1)(\frac{h}{2}s+1)}{(\lambda s+1)^2}\frac{1}{s}\right\} \tag{2.57}$$

where $\mathcal{L}^{-1}$ denotes the inverse Laplace transform. By calculating (2.57) the maximum value for $u(t)$ can be found to be

$$\|u(t)\|_\infty = \begin{cases} \frac{1}{K_g}\left(1 - \frac{\tau h - 2\tau\lambda - h\lambda + 2\lambda^2}{2\lambda^2}e^{-\frac{2\tau h - 2\tau\lambda - h\lambda}{(-2\lambda+h)(\tau-\lambda)}}\right) & \text{if } \tau > \lambda \\ \frac{1}{K_g} & \text{if } \tau \le \lambda \end{cases} \tag{2.58}$$

From this point on, it is assumed that $\lambda/h > 0.5$ (indeed, it was shown in Figure 2.7 that $\lambda/h \ge 0.6$ for zero nominal overshoot). From (2.58) it can be seen that $\|K_g u(t)\|_\infty$ only depends on $\frac{h}{\tau}$ and $\frac{\lambda}{h}$. Thus, once a given plant has been modelled according to a FOPTD description, a particular $\frac{h}{\tau}$ relation is obtained. With this in mind, $\|K_g u(t)\|_\infty$ only depends on $\frac{\lambda}{h}$, and $\lambda$ can be tuned so as to obtain acceptable peaks on the control signal. On the other hand, if $\frac{\lambda}{h}$ is fixed first, then $\|K_g u(t)\|_\infty$ becomes a function of $\frac{h}{\tau}$. This function can be used to give an idea of the maximum control effort required along different *lead-lag* ratios. In our case, we fixed $\frac{\lambda}{h} = 1$. Substituting $\lambda = h$ in (2.58) leads to

$$\|u(t)\|_\infty = \begin{cases} \frac{1}{K_g}\left(1 - \frac{1}{2}\left(\frac{-\tau+h}{h}\right)e^{\frac{h}{-\tau+h}}\right) & \text{if } h < \tau \\ \frac{1}{K_g} & \text{if } h \ge \tau \end{cases} \tag{2.59}$$

which can be easily expressed as a function of $K_g$ and $\frac{h}{\tau}$.

Slew rate constraints can be similarly tackled. From (2.57) the derivative of the control signal is

$$\dot{u}(t) = \mathcal{L}^{-1}\{Q\} = \mathcal{L}^{-1}\left\{\frac{1}{K_g}\frac{(\tau s + 1)(\frac{h}{2}s + 1)}{(\lambda s + 1)^2}\right\} \tag{2.60}$$

By calculating (2.60) and considering the following decomposition[4]

$$\dot{u}(t) = \dot{u}_+(t) + \dot{u}_-(t) \tag{2.61}$$

where

$$\dot{u}_+(t) \doteq \left\{\begin{array}{ll} \dot{u}(t) & \text{if} \quad \dot{u}(t) > 0 \\ 0 & \text{if} \quad \dot{u}(t) \le 0 \end{array}\right. \quad \text{and} \quad \dot{u}_-(t) \doteq \left\{\begin{array}{ll} -\dot{u}(t) & \text{if} \quad \dot{u}(t) < 0 \\ 0 & \text{if} \quad \dot{u}(t) \ge 0 \end{array}\right. \tag{2.62}$$

the following expressions can be easily obtained

$$\|\dot{u}_+(t)\|_\infty = -\frac{1}{2}\frac{2\tau h - 2\tau\lambda - h\lambda}{K_g\lambda^3} \tag{2.63}$$

and

$$\|\dot{u}_-(t)\|_\infty = \left\{\begin{array}{ll} \frac{1}{2}\frac{2\lambda^2 - 2\tau\lambda - h\lambda + \tau h}{K_g\lambda^3}e^{-\frac{3\tau h + 2\lambda^2 - 4\tau\lambda - 2h\lambda}{(-2\lambda + h)(\tau - \lambda)}} & \text{if} \quad \tau > \lambda \\ 0 & \text{if} \quad \tau \le \lambda \end{array}\right. \tag{2.64}$$

which represent, respectively, the highest rates of change in the increasing and decreasing directions. For the proposed tuning rule $\lambda = h$ the above two expressions simplify to

$$\|\dot{u}_+(t)\|_\infty = \frac{1}{2K_g h} \tag{2.65}$$

and

$$\|\dot{u}_-(t)\|_\infty = \left\{\begin{array}{ll} \frac{1}{2}\frac{h - \tau}{K_g h^2}e^{-\frac{\tau}{\tau - h}} & \text{if} \quad h < \tau \\ 0 & \text{if} \quad h \ge \tau \end{array}\right. \tag{2.66}$$

## 2.5 Simulation examples

In this section we will evaluate the proposed simple automatic tuning rule of Table 2.1 through simulations. The objective is to cover a representative set of examples so as to properly obtain conclusions regarding the performance and robustness of the suggested method. Table 2.3 collects the information of the experimental setup. Firstly, four linear processes are considered including the lag-dominant, lead-dominant and balanced lag and delay cases. The first one consists of a FOPTD plant for which there is only parametric uncertainty whereas the other three systems are linear processes modelled as FOPTD plants. These three last examples are taken from [19]. Additionally, a fifth nonlinear system is taken into account. This last process represents the isothermal series/paralel Van de Vusse reaction [79] taking place in an isothermal continuous stirred tank reactor (CSTR). The corresponding approximate FOPTD model has been derived assuming the system in a stationary point.

For the sake of comparison, other approaches to PID design considering FOPTD models are examined. Since a complete comparison is not possible due to the large number of existing tuning rules (see [51]) we will concentrate on two existing methods also conceived in the spirit of simplicity:

- SIMC tuning rule (leading to a PI). A simple and effective tuning proposed in [69].

---

[4]A similar decomposition was not used for $u(t)$ in (2.57) due to the fact that $u(t) \ge 0$ for $t \ge 0$.

**TABLE 2.3**

Processes within the experimental setup together with their FOPTD approximations. $P_{1-4}$ are linear processes. Regarding $P_5$, $\mathbf{f}(\mathbf{x}, u) = (f_1(\mathbf{x}, u), f_2(\mathbf{x}, u)) = (-50x_1 - 10x_1^2 + (10 - x_1)u, 50x_1 - 100x_2 - x_2 u)$. The assumed working point is $(\mathbf{x}^*, u^*) = (3, 1.117, 34.2805)$.

| Real Process | FOPTD Model |
|---|---|
| $P_1 = \dfrac{1.2e^{-1.2s}}{0.8s+1}$ | $\dfrac{e^{-s}}{s+1}$ |
| $P_2 = \dfrac{1}{(1+s)(1+0.1s)(1+0.01s)(1+0.001s)}$ | $\dfrac{e^{-0.073s}}{1.073s+1}$ |
| $P_3 = \dfrac{e^{-s}}{(1+0.05s)^2}$ | $\dfrac{e^{-s}}{0.093s+1}$ |
| $P_4 = \dfrac{1}{(1+s)^4}$ | $\dfrac{e^{-1.42s}}{2.9s+1}$ |
| $P_5 \equiv \begin{cases} \dot{\mathbf{x}} &= \mathbf{f}(\mathbf{x}, u) \\ y &= x_2 \end{cases}$ | $\dfrac{0.0126e^{-0.0085s}}{0.01s+1}$ |

- AMIGO tuning rule (leading to a PID). A rule along the lines of the classical Ziegler-Nichols method. See [19].

In order to evaluate the robustness and the performance obtained with the different methods at hand, the following standard measures will be used:

- *Robustness*: The peak of the sensitivity function, $M_S$, is the inverse of the minimum distance from the Nyquist plot to the critial point and constitutes a quite standard robustness indicator [72].

- *Output performance*: The integrated absolute error (IAE) of the error $e = r - y$ will be computed.

$$\text{IAE} = \int_0^\infty |e(t)| dt$$

- *Input performance*: To evaluate the manipulated input usage, the total variation (TV) of the control signal $u(t)$ will be computed.

$$\text{TV} = \int_0^\infty |\dot{u}(t)| dt$$

To provide a more global and complete comparison framework, the performance measures above will be calculated for both a set-point change and load disturbance. In addition, the percent overshoot of the output $y(t)$, denoted by $y_{ov}$, will be taken into account for set-point output performance. Similarly, the peak of the control signal $u(t)$ (i.e., $\|u(t)\|_\infty$) will be indicated for load disturbance performance. Table 2.4 summarizes the results obtained.

It follows from Figures 2.10–2.14 that the proposed tuning rule generates quite smooth responses requiring a moderate control action level.

Table 2.4 shows that the required control usage is lower than that associated with the other two methods. In particular, for plants $P_1$ and $P_5$ the proposed tuning control usage is far below that of the AMIGO tuning. It can also be seen that the proposed method provides,

**TABLE 2.4**

Results of performance/robustness evaluation for the set of plants $\{P_i\}_{i=1}^5$. As the system $P_5$ is nonlinear, the robustness indicator $M_S$ has been computed with respect to its linearization on the working point, which turns out to be $\frac{-1.117s+188.8}{s^2+278.6s+1.937e04}$.

| Plant | Tuning | Robustness $M_S$ | Performance set-point IAE | TV | $y_{ov}$ | disturbance IAE | TV | $\|u(t)\|_\infty$ |
|-------|--------|------|------|------|------|------|------|------|
| $P_1$ | Proposed | 1.74 | 2.15 | 1.1 | 3.3 | 1.25 | 0.55 | 0.51 |
|  | SIMC | 2.12 | 2.4 | 1.64 | 20.95 | 1.235 | 0.79 | 0.6 |
|  | AMIGO | 1.85 | 1.7 | 6.13 | 10.96 | 0.91 | 0.72 | 0.56 |
| $P_2$ | Proposed | 1.33 | 0.21 | 11.03 | 6.36 | 0.09 | 0.57 | 0.53 |
|  | SIMC | 1.56 | 0.24 | 16.11 | 23.42 | 0.04 | 0.75 | 0.62 |
|  | AMIGO | 1.31 | 0.24 | 11.56 | 23.55 | 0.027 | 0.75 | 0.62 |
| $P_3$ | Proposed | 1.42 | 2.5 | 1 | 0 | 1.25 | 0.5 | 0.5 |
|  | SIMC | 1.6 | 2.18 | 1.09 | 4.27 | 1.09 | 0.54 | 0.52 |
|  | AMIGO | 1.46 | 1.94 | 1.46 | 0 | 0.97 | 0.56 | 0.5 |
| $P_4$ | Proposed | 1.66 | 3.76 | 1.78 | 5.1 | 1.77 | 0.61 | 0.53 |
|  | SIMC | 2 | 4.08 | 2.68 | 18.77 | 1.54 | 0.82 | 0.6 |
|  | AMIGO | 1.62 | 3.23 | 2.02 | 15.54 | 1.26 | 0.67 | 0.58 |
| $P_5$ | Proposed | 1.4 | 0.0026 | 11.1 | 0.4 | 0.0005 | 2.3 | 34.46 |
|  | SIMC | 1.57 | 0.0026 | 15.6 | 1.6 | 0.0047 | 2.5 | 34.48 |
|  | AMIGO | 1.38 | 0.0022 | 51.4 | 1.1 | 0.0041 | 3 | 34.73 |

generally, both the minimum output overshoots and the control signal peaks. With respect to set-point evaluation, the AMIGO tuning rule gives better IAEs. However, if one inspects Figures 2.10–2.14, it is clear that the set-point responses of the proposed method exhibit less overshoot than those of the AMIGO tuning.

Regarding disturbance rejection, the proposed method provides an inferior performance with respect to the SIMC and AMIGO proposals. This is a quite expected result since the proposed tuning rule was derived for smooth set-point. Nevertheless, disregarding the lag-dominant plant $P_2$, the disturbance rejection responses are not significantly inferior, and in the case of the SIMC, they are indeed quite comparable. This is explained, in part, by the fact that the optimization problem in (2.31) is, as a matter of fact, a sensitivity optimization problem. A deeper analysis of the *servo/regulation* trade-off tuning within a generalized version of the presented framework is needed, and it will be presented in the next chapters.

Lastly, Table 2.4 shows that for the five considered systems the robustness indicator for the proposed method is always very close to that associated with the best method. This robustness is in accordance with the smoothness of the corresponding control and output signals.

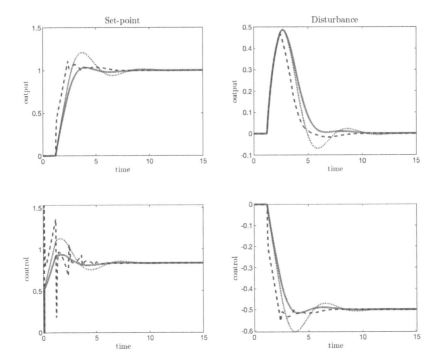

**FIGURE 2.10**
$P_1$ time responses for set-point change and load disturbance for the proposed (solid), SIMC (dotted), AMIGO (dashed), and [81] (dashdot) tuning rules.

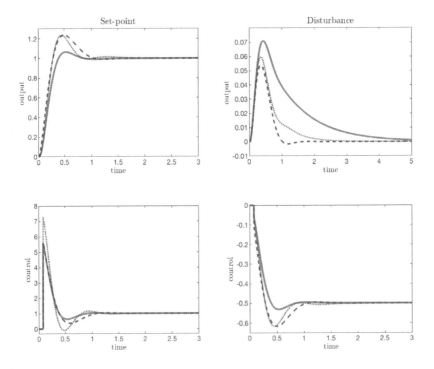

**FIGURE 2.11**
$P_2$ time responses for set-point change and load disturbance for the proposed (solid), SIMC (dotted), AMIGO (dashed), and [81] (dashdot) tuning rules.

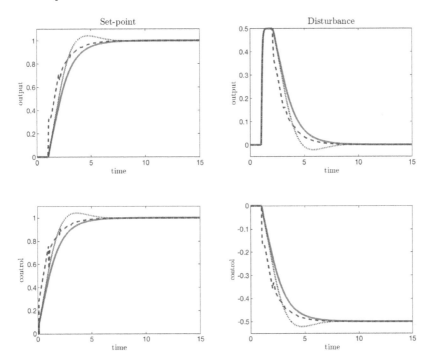

**FIGURE 2.12**
$P_3$ time responses for set-point change and load disturbance for the proposed (solid), SIMC (dotted), AMIGO (dashed), and [81] (dashdot) tuning rules.

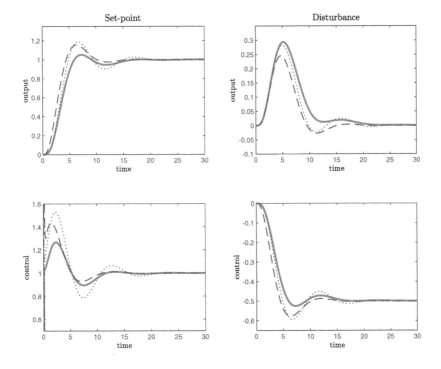

**FIGURE 2.13**
$P_4$ time responses for set-point change and load disturbance for the proposed (solid), SIMC (dotted), AMIGO (dashed), and [81] (dashdot) tuning rules.

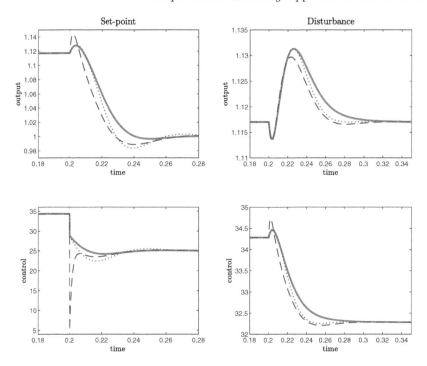

**FIGURE 2.14**
$P_5$ time responses for set-point change and load disturbance for the proposed (solid), SIMC (dotted), AMIGO (dashed), and [81] (dashdot) tuning rules.

# 3

## Alternative Design for Load Disturbance Improvement

Due to the fact that sometimes a one-degree-of-freedom (1DOF) PID has to deal with both set-point changes and load disturbances, it would be desirable to have at one's disposal simple methods to achieve a good compromise in this situation. Based on the well-established min-max model matching theory, this chapter addresses the smooth tuning of a 1DOF PID controller for both acceptable load disturbance attenuation and set-point tracking. As the design specifications are commonly given in the form of maximum allowed overshoot, peak on the sensitivity function, and other popular measures, the analysis to carry out the design quantitatively based on these indexes is also provided.

This chapter is organized as follows: Section 3.1 is devoted to the problem statement. Section 3.2 reviews the model-matching analytical $\gamma$-tuning design. The exact interval for $\gamma$ considering load disturbances is determined in Section 3.3, which also concerns the stability analysis. In Section 3.4, the quantitative tuning of $\gamma$ is addressed according to standard robustness/performance indexes. Section 3.5 illustrates with some examples the suggested methodology for balancing servo and regulator performance.

## 3.1 Problem statement

It is well-known that today most of the control systems in industry are still operated by PID controllers [20, 66]. This is somehow explained due to their simplicity and acceptable performance in practice. Consequently, given the widespread use of PID compensators, it is clear that even a small improvement over the already existing tuning methods could represent a benefit in process control. Fortunately, the gap between the theory and practice in control engineering has been reduced during the last twenty years in an attempt to incorporate the methods of optimal and robust control theory to the PID area. As a consequence, new theoretical results, as well as several PID designs, have been reported [32, 69, 19, 81, 33, 62], confirming a trend that endures.

Among the well-established analytical methods, the design of compensators by means of a desired closed loop specification is a quite common one. Regarding PID controllers, some model-matching–based approaches have appeared during the last years [1, 81]. The problem with the approach in [1] is that robustness is not considered explicitly and no tuning rule is finally provided. Avoiding these problems, in the previous chapter an analytical model-matching–based robust PID design is proposed for smooth set-point tracking. In this chapter, we extend this approach so that the controller can be tuned somewhere in between the *servo* and the *regulator* operating modes by means of adjusting a single design parameter: $\gamma$. Obviously, within the 1DOF context it is not possible to have both optimal set-point and load disturbance attenuation simultaneously [72]. Thus, some kind of trade-off is necessary in order to minimize the overall performance degradation. From a numerical point of view, the underlying idea was originally presented in [16], and further developed in [17]. The contributions of this chapter are summarized below:

- First, the interval for the $\gamma$ parameter in [82] is determined considering that the disturbances enter at the input of the plant (i.e., load disturbances). Additionally, the nominal stability analysis is conducted.

- Second, once the interval for $\gamma$ is known, quantitative tuning guidelines based on common robustness/performance indexes are given.

In [20], it is claimed that many analytical methods for PID design produce pole-zero cancellations, which makes them unsuitable for regulator purposes. This is one of the drawbacks of the conventional IMC method [49, 69] and some IMC-like approaches [92, 81, 13, 8]. These approaches are suitable for servo operation but may exhibit poor disturbance attenuation. On the contrary, the method presented here is aimed at providing good responses in both servo and regulator mode.

### 3.1.1    The control framework

The conventional 1DOF scenario of Figure 3.1 is assumed, where a distinction is made between input ($d_i$) and output ($d_o$) disturbances. A FOPTD model for the plant $P$ is used, i.e.:

$$P = K_g \frac{e^{-sh}}{\tau s + 1} \tag{3.1}$$

For design purposes, it is convenient to approximate the delay term in (3.1) so as to achieve a purely rational process model. By using the first-order Taylor expansion $e^{-sh} \approx 1 - sh$ , the model in (3.1) can be approximated as follows

$$P \approx K_g \frac{-sh + 1}{\tau s + 1} \tag{3.2}$$

Regarding the control law, as in the previous chapter, the ISA PID form [20] is considered:

$$u = K_c \left(1 + \frac{1}{sT_i} + \frac{sT_d}{1 + sT_d/N}\right) e \tag{3.3}$$

where $e = r - y$, being $r$, $y$, and $u$ the Laplace transforms of the reference, process output and control signal, respectively. $K_c$ is the PID gain, whereas $T_i$ and $T_d$ are its integral and derivative time constants, respectively. Finally, $N$ is the ratio between $T_d$ and the time constant of an additional pole introduced to assure the properness of the

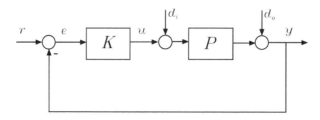

**FIGURE 3.1**
Feedback control scheme.

controller. Therefore, the transfer function representation of the chosen controller $K$ is given by

$$K = K_c \frac{1 + s(T_i + \frac{T_d}{N}) + s^2 T_i \frac{T_d}{N}(N+1)}{s T_i (1 + s\frac{T_d}{N})} \tag{3.4}$$

### 3.1.2   The model-matching problem formulation

In this section, the min-max model matching approach to controller design is presented. The controller derivation will be based on a desired model for a given closed loop transfer function. Mathematically, a min-max optimization problem of the form

$$\min_{K \in \mathcal{C}} \left\| W(X_d - X) \right\|_\infty \tag{3.5}$$

is posed to capture the performance objective. $W$ is a weighting function, $X$ is a given closed loop system relation, and $X_d$ is the target model for $X$. For instance, one could choose $X = T$, the complementary sensitivity function. In this case, $X_d = T_d$ would specify a desired input-to-output response for the closed loop system. Another possibility is to choose $X = S$, the sensitivity function. In this case, $X_d = S_d$ would represent the desired sensitivity shape in frequency. Note that

$$\min_{K \in \mathcal{C}} \left\| \mathcal{E} \right\|_\infty = \min_{K \in \mathcal{C}} \left\| W(T_d - T) \right\|_\infty \tag{3.6}$$

$$= \min_{K \in \mathcal{C}} \left\| W((1 - S_d) - (1 - S)) \right\|_\infty$$

$$= \min_{K \in \mathcal{C}} \left\| W(S_d - S) \right\|_\infty \tag{3.7}$$

Consequently, (3.6) and (3.7) ultimately amount to the same optimization problem. However, depending on the choice of the desired relations, the optimization will be better regarded as a complementary sensitivity problem for set-point tracking purposes or as a sensitivity one for disturbance rejection. By using the Youla parameterization (2.24), the constrained problem (3.6) can be recast in the more convenient form:

$$\min_{Q \in \mathcal{RH}_\infty} \left\| \mathcal{E} \right\|_\infty = \min_{Q \in \mathcal{RH}_\infty} \left\| W(T_d - PQ) \right\|_\infty \tag{3.8}$$

## 3.2   Model-matching solution for PID design

With the purpose of making this current work more self-contained, this section briefly outlines the PID-oriented solution to (3.8) presented in [81, 82]. In order to constrain the solution of (3.8) to be a PID compensator, the following setting was proposed:

$$T_d = \frac{(T_M - \gamma)s + 1}{1 + T_M s} \tag{3.9}$$

which corresponds to a desired sensitivity function $S_d$ of the form

$$S_d = \frac{\gamma s}{T_M s + 1} \tag{3.10}$$

with respect to the sensitivity problem (3.7). Regarding the weighting function in (3.8), the following choice was made:

$$W = \frac{1 + zs}{s} \tag{3.11}$$

in order to automatically include integral action and keep it as simple as possible. The solution to the MMP (3.8) can be obtained applying Lemma 2.1.1 with $T_1 = T_d$ and $T_2 = PQ$, where $P$ is taken as in (3.2). In this case, there is only one RHP zero in $P$ ($\nu = 1$) at $s = \frac{1}{h}$. Consequently, the optimum error is given by $\mathcal{E}^o = \rho$, being $\rho$ determined from a single interpolation constraint:

$$\mathcal{E}^o(1/h) = \rho = W(1/h)T_d(1/h) \implies \rho = \frac{1 + \frac{z}{h}}{\frac{1}{h}} \frac{(T_M - \gamma)\frac{1}{h} + 1}{1 + \frac{T_M}{h}} \tag{3.12}$$

or more compactly:

$$\rho = \frac{h + z}{h + T_M}(T_M + h - \gamma) \tag{3.13}$$

Now, the optimum $Q$ can be easily computed

$$\rho = \mathcal{E}^o = W(T_d - PQ) \implies Q = \left(T_d - \rho W^{-1}\right)P^{-1} \tag{3.14}$$

By taking $P$ as in (3.2) and substituting (3.9), (3.11), and (3.13) into (3.14), the optimal $Q$ is

$$Q = \frac{1}{K_g}\frac{(1 + \tau s)(1 + \chi s)}{(1 + T_M s)(1 + zs)} \tag{3.15}$$

with

$$\chi = z + h - \rho + T_M - \gamma \tag{3.16}$$

The equivalent unity feedback controller is given by (6.4):

$$K = \frac{1}{K_g(\rho + \gamma)}\frac{(1 + \tau s)(1 + \chi s)}{s(1 + \frac{zT_M + h\chi}{\rho + \gamma}s)} \tag{3.17}$$

The compensator in (3.17) can be cast into (3.4) in accordance with:

$$\begin{aligned} K_c &= \frac{T_i}{K_g(\rho + \gamma)} \\ T_i &= \tau + \chi - \frac{zT_M + h\chi}{\rho + \gamma} \\ \frac{T_d}{N} &= \frac{zT_M + h\chi}{\rho + \gamma} \\ N + 1 &= \frac{\tau}{T_i}\chi\frac{\rho + \gamma}{zT_M + h\chi} \end{aligned} \tag{3.18}$$

It can be seen that there are three design parameters: $T_M$, $z$ and $\gamma$. The meaning of $T_M$ and $z$ can be easily understood by considering the choice $\gamma = T_M$. This way, $\gamma$ disappears, and $T_d$ in (3.9) becomes

$$T_d = \frac{1}{1 + T_M s} \tag{3.19}$$

Accordingly, (3.8) represents now a MMP with respect to the desired input-to-output relation $T_d$ in (3.19). Based on this particular scenario, $z$ and $T_M$ can be fixed to obtain smooth set-point responses [81]. The role of $T_M$ is clear: it captures the desired closed loop bandwidth. The role of $z$ is to provide tolerance to model uncertainty. Let us assume first that the dynamic behaviour of the plant under control is described not only by the nominal model but by a whole family of possible plants:

$$\mathcal{F} = \left\{\tilde{P} = P(1 + \Delta_m)\right\} \tag{3.20}$$

where $\Delta_m$ is the relative (multiplicative) model error

$$\Delta_m \doteq \frac{\tilde{P} - P}{P} \tag{3.21}$$

satisfying $|\Delta_m(j\omega)| \leq |W_m(j\omega)|$ and $W_m$ is a frequency weight bounding the modelling error. It is well-known [72, 49] that a controller $K$ that stabilizes the nominal plant $P$, also stabilizes all the plants in (3.20) provided that

$$\|W_m T\|_\infty < 1 \tag{3.22}$$

From (3.2), (3.15), and the fact that $T = PQ$, the robust stability condition (3.22) can be written as

$$\left| \frac{(1 - sh)(1 + \chi_1 s)}{(1 + T_m j\omega)(1 + zj\omega)} \right| < \left| \frac{1}{W_m(j\omega)} \right| \quad \forall \omega \tag{3.23}$$

where $\chi_1 = z + \tau - \rho$. Let us consider $|W_m(j\omega)| = 1$ in (3.23), giving rise to the following condition:

$$\left| \frac{(1 - sh)(1 + \chi_1 s)}{(1 + T_m j\omega)(1 + zj\omega)} \right| < 1 \quad \forall \omega \tag{3.24}$$

Although this may seem conservative, mid- and high-frequency uncertainty will be allowed. The constraint (3.24) imposes that the closed loop transfer function does not have *spikes* and behaves as a low-pass filter. By choosing

$$T_M = \sqrt{2}h \qquad z = 2h \tag{3.25}$$

a suitable pole-zero pattern is obtained, leading to a complementary sensitivity function with the desired low-pass shape. Selecting $T_M$ and $z$ as in (3.25) (with $\gamma = \gamma_{sp} = T_M$) yields a smooth set-point response. Nevertheless, as it was pointed out in [81], this design can lead to poor disturbance rejection. Improving the disturbance rejection is precisely the role of the $\gamma$ parameter.

In [82] it is shown that by choosing $\gamma > T_M = \sqrt{2}h$, the disturbance attenuation can be improved. However, the analysis in [82] only takes into account disturbances entering at the output of the plant, remaining open in how to select $\gamma$ if the disturbance enters at the input. Another missing aspect is the stability analysis. These points are the subject matter of the next subsection.

**Remark 3.2.1.** *In what follows, $z$ and $T_M$ will remain fixed as indicated in (3.25). Therefore, the tuning rule (1.20) depends on a single tuning parameter: $\gamma$. As will be illustrated in the next subsection, $\gamma = \gamma_{sp} = \sqrt{2}h$ corresponds to the servo operation. From this point on, increasing the value of $\gamma$ allows improvement of the regulatory performance. In this sense, (1.20) can be considered as a (robust) unified servo/regulation tuning rule.*

## 3.3   Trade-off tuning interval considering load disturbances

First, we will review how to select $\gamma$ for optimal step disturbance rejection at the output of the plant. Assuming that $P = \tilde{P}$ in Figure 2.2, we have that $y = (1 - PQ)d_o$. Taking $d_o = \frac{1}{s}$, $P$ as in (3.2) and $Q$ from (3.15) finally gives

$$y = S\frac{1}{s} \approx \left( 1 - \frac{(1 - sh)(1 + (\sqrt{2}h + 0.24\gamma) s)}{(1 + \sqrt{2}hs)(1 + 2hs)} \right) \frac{1}{s} \tag{3.26}$$

We can choose $\gamma$ according to the integral squared error, defined as

$$ISE \doteq \int_0^\infty \big(r(t) - y(t)\big)^2 \, dt \tag{3.27}$$

As we are concerned with the output disturbance rejection, we can assume that the reference is zero ($r(t) = 0$), which leads to

$$ISE(\gamma, h) = \int_0^\infty y^2(t)dt = \|y\|_2^2 \tag{3.28}$$

By applying Parseval's theorem [49], this calculation can be rewritten as

$$ISE(\gamma, h) = \frac{1}{2\pi} \int_{-\infty}^{\infty} y(j\omega)y(-j\omega)d\omega = \frac{1}{2\pi j} \oint y(s)y(-s)ds \tag{3.29}$$

Applying of the residue theorem [24] to solve (3.29) yields

$$ISE(\gamma, h) \approx 2.25h - 0.1065\gamma + \frac{0.0112}{h}\gamma^2 \tag{3.30}$$

By taking the derivative with respect to $\gamma$ we can obtain the optimal value that minimizes the ISE criterion

$$\frac{\partial ISE(\gamma, h)}{\partial \gamma} = 0 \Rightarrow \gamma_{ldo} \approx 4.56h \tag{3.31}$$

In [82] it was shown that if the disturbance occurs at the input of the plant, the disturbance rejection produced by taking $\gamma$ as in (3.31) may still be improved significantly, but the exact tuning was not given. In what follows, we will address how to choose $\gamma$ for disturbances at the input of the plant. In this situation, the disturbance to output relation is

$$y = PSd_i \tag{3.32}$$

Instead of optimizing the ISE criterion as before, a more *heuristic* approach is taken. In the *lag-dominant* case, it is evident that a sluggish response will be obtained unless $S$ cancels the slow dynamic of $P$. In fact, the impossibility of producing such a cancellation is the reason why the IMC-like design in [13] cannot be used for regulator purposes. Thus, it would be necessary for good input disturbance attenuation that

$$S|_{s=-\frac{1}{\tau}} = 0 \tag{3.33}$$

Taking $S$ as in (3.26), the following value for $\gamma$ is finally obtained

$$\gamma_{ldi} \approx -\frac{12.36h(-\tau + \sqrt{2}h)}{h + \tau} \tag{3.34}$$

It is clear that $\gamma_{ldi}$ in (3.34) can be regarded as a function of just $\frac{\tau}{h}$. Provided that $\tau \geq 3h$, the following chain of inequalities holds

$$\gamma_{sp} = \sqrt{2}h < \gamma_{ldo} \approx 4.56h < \gamma_{ldi} \tag{3.35}$$

For example, in the case $\tau/h = 10$, it turns out that $\gamma_{ldi} \approx 9.65h$. We will assume from this point on that disturbances enter at the input of the plant. Thus, we will tune $\gamma$ as in (3.34) for the regulator mode. Consequently, we are finally considering the following interval for the $\gamma$ parameter:

$$\gamma \in [\gamma_{sp} = \sqrt{2}h, \gamma_{ld} = \gamma_{ldi}] \tag{3.36}$$

where the extremes represent the tuning for servo ($\gamma = \gamma_{sp}$) and regulator ($\gamma = \gamma_{ld}$) operation.

### 3.3.1   Nominal stability

Since we have considered the approximation (3.2) for the FOPTD model, the basic requirement of nominal stability is not automatically guaranteed for a FOPTD plant even when all its parameters are perfectly known. The nominal stability issue is dealt with here by means of the dual locus technique along the lines of [94]. Taking the unity feedback controller from (3.17) together with the model (3.1) results in the following loop transfer function

$$L \approx \frac{1 + \left( \sqrt{2}h + 0.24\gamma \right) s}{(3h - 0.24\gamma) s + \left( 3\sqrt{2}h^2 + 0.24h\gamma \right) s^2} e^{-sh} \tag{3.37}$$

The characteristic equation $1 + L = 0$ can then be rewritten in the form

$$L_1 - L_2 = 0 \tag{3.38}$$

by making the following assignments:

$$L_1 = -\frac{(3h - 0.24\gamma) s + \left( 3\sqrt{2}h^2 + 0.24h\gamma \right) s^2}{1 + \left( \sqrt{2}h + 0.24\gamma \right) s} \qquad L_2 = e^{-sh}$$

The dual locus diagram technique states that the closed loop system is stable if the Nyquist locus of $L_1$ reaches the intersection point with the locus of $L_2$ before $L_2$ does so. In this situation, the number of encirclements of $L_1 - L_2$ around the origin when traversing the Nyquist contour is zero. In our case, $L$ has a pole at the origin and the conventional Nyquist contour must be modified to avoid passing through the point $0 + j0$ when traversing the imaginary axis. One way to do it is to construct a semicircular arc with infinitely small radius $r$ around $0 + j0$, that starts at $0 + j(0 - r)$ and travels counterclockwise to $0 + j(0 + r)$. For our purposes, however, it is enough to consider the Nyquist contour on the positive imaginary axis. Thus, by the argument principle [24], there is no RHP pole in the closed loop system. The idea is illustrated for a particular case in Figure 3.2. The phase angles of $L_1$ and $L_2$ at $\omega_c$ (the intersection frequency) are, respectively:

$$\phi_1 = \arctan \frac{-3h + 0.24\gamma}{\left( 3\sqrt{2}h^2 + 0.24h\gamma \right) \omega_c} - \arctan \left( \sqrt{2}h + 0.24\gamma \right) \omega_c \tag{3.39}$$

and

$$\phi_2 = -h\omega_c \tag{3.40}$$

The stability condition is satisfied only when the phase angle of $L_1$ is larger (in absolute value) than that of $L_2$, i.e., if $\phi_1 - \phi_2 < 0$ (see Figure 3.2). Since $|L_2(j\omega)| = 1$, the intersection frequency can be determined from

$$\left| -\frac{(3h - 0.24\gamma) s + \left( 3\sqrt{2}h^2 + 0.24h\gamma \right) s^2}{1 + \left( \sqrt{2}h + 0.24\gamma \right) s} \right|_{s=j\omega_c} = 1 \tag{3.41}$$

Solving (3.41) leads to

$$\omega_c = \frac{1}{h \left( 6 + 0.24\sqrt{2}\eta \right)} \sqrt{-(B^2 - C^2)^2 + \sqrt{(B^2 - C^2)^2 + 4A^2}} \tag{3.42}$$

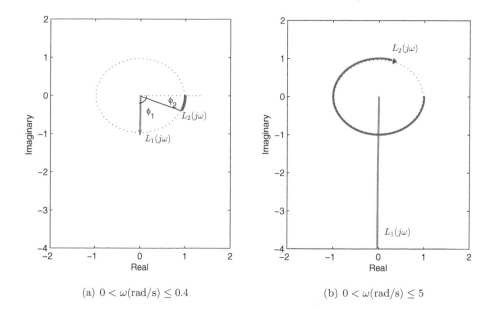

(a) $0 < \omega(\text{rad/s}) \leq 0.4$                    (b) $0 < \omega(\text{rad/s}) \leq 5$

**FIGURE 3.2**
Dual locus diagram (restricted to the positive imaginary axis only) for $h = 1, \gamma = \sqrt{2}$. The closed loop is stable because $\phi_1 - \phi_2 < 0$ at the intersection frequency $\omega_c$, approximately 0.4 rad/s.

where

$$
\begin{aligned}
A &= 3\sqrt{2} + 0.24\eta \\
B &= 3 - 0.24\eta \\
C &= \sqrt{2} + 0.24\eta
\end{aligned}
\tag{3.43}
$$

and $\eta \doteq \frac{\gamma}{h}$. It can be seen from (3.42) and (3.43) that $\omega_c h$ can be expressed as a function of $\frac{\gamma}{h}$. From (3.39) and (3.40), this in turn shows that $\phi_1 - \phi_2$ depends solely on $\frac{\gamma}{h}$. As we are considering the interval in (3.36), then $\frac{\gamma}{h} \in [\sqrt{2}, \frac{12.36(\tau/h - \sqrt{2})}{1 + \tau/h}]$. The maximal interval corresponds to the situation in which $\tau/h \to \infty$, which yields $\sqrt{2} \leq \frac{\gamma}{h} < 12.36$. In Figure 3.3, $\phi_1 - \phi_2$ is plotted against $\frac{\gamma}{h}$. It can be concluded that the closed loop system remains stable within the considered maximal interval. Hence, the nominal stability is always guaranteed.

So far, we have verified that the control system is stable with respect to the FOPTD model. Sometimes, however, the available model of the plant does not correspond to a FOPTD process. In these cases, we adopt a simple FOPTD model for the sake of simplicity: a low-order model will yield a low-order controller. Consequently, in theses cases, the nominal stability must consider the mismatch between the approximate FOPTD model (3.2) and the one precisely describing the dynamics of the real process. If we assume that the real plant $\tilde{P}$ is perfectly known, we can take $\Delta_m$ as in (3.21) and use the robust stability condition (3.22) for nominal stability purposes. As in this specific case, we are considering that $\Delta_m$ is perfectly known (we know both $\tilde{P}$ and $P$). We can apply (3.22) taking $W_m = \Delta_m$ and restricted to the phase crossover frequency only:

$$
|T(j\omega_{180})\Delta_m(j\omega_{180})| < 1
\tag{3.44}
$$

where $\omega_{180}$ verifies

$$
\angle \left( T(j\omega_{180})\Delta(j\omega_{180}) \right) = -180°
\tag{3.45}
$$

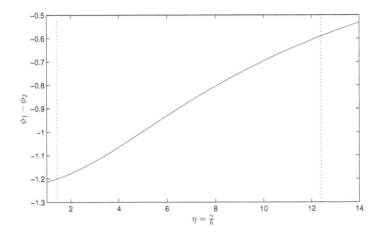

**FIGURE 3.3**
The closed loop is stable because $(\phi_1 - \phi_2)(\eta) < 0$.

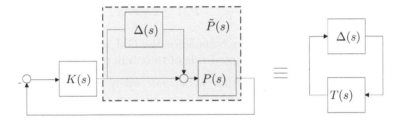

**FIGURE 3.4**
Equivalent stability loops.

This can be readily understood by looking at Figure 3.4: the first feedback loop is stable if and only if the second one is stable. As $T\Delta = PQ\frac{\tilde{P}-P}{P} = Q(\tilde{P} - P)$ is stable (we are considering stable plants), condition (3.44) coincides with the Nyquist stability criterion [72] when $\angle(T(j\omega)\Delta(j\omega))$ crosses $-180°$ only once. If it crosses $-180°$ several times (this would seldom happen in practice), then (3.44) must account for all the phase crossover frequencies and becomes only a sufficient condition.

In the next section, guidelines for the tuning of $\gamma$ are given. In particular, robustness is captured in terms of the peak on the sensitivity function.

## 3.4  Tuning guidelines

In classical control theory, the performance of a control system is usually characterized in terms of the transient and steady-state time-domain responses. Well-known time-domain performance indicators are, among others, the overshoot and the settling time. On the other hand, the classical frequency domain approach provides typical robustness indexes as, for example, the GM (gain margin), the PM (phase margin), or the peak on the sensitivity function ($M_S$). These and other common standard measures usually capture the design specifications in practical designs and have something in common: they all depend only on the finally achieved loop function $L \doteq PK$. This is precisely why one of the earliest design strategies was the *loop-shaping* method, which consisted of designing the feedback compensator indirectly through a desired loop function meeting the control requirements.

This section is aimed at showing that the optimization procedure used to derive the presented controller can be ultimately put in connection with the classical robustness/performance indicators. Only the overshoot and the $M_S$ specifications will be considered, for which the following qualitative facts hold:

- The greater the value of $\gamma$, the greater the overshoot. This point is important because some plants have a strict limitation on the allowable overshoot.

- Augmenting $\gamma$ also reduces the stability margins, i.e., augments $M_S$. Recall that $M_S$, defined as

$$M_S \doteq \|S(j\omega)\|_\infty \doteq \max_\omega \left| \frac{1}{1 + L(j\omega)} \right|,$$

corresponds to the inverse of the shortest distance from the Nyquist plot to the critical point. This constitutes a more reliable robustness indicator than the popular gain and phase margins. This is because a control system with good gain and phase margins can be very close to instability. On the contrary, it can be seen [72] that

$$GM \geq \frac{M_S}{M_S - 1} \qquad PM \geq 2 \arcsin\left(\frac{1}{2M_S}\right)$$

In concrete, $M_s = 2$ provides lower bounds for the GM and PM of 2 and 30 degrees, respectively, which are a quite accepted rule of thumb in practice.

From (3.37), it is easily observed that $L$ depends solely on $h$ and $\gamma$ in the nominal case. This means that the classical measures depending on $L$ can be finally expressed as functions of the variables $h$ and $\gamma$. Moreover, by applying the Buckingham pi theorem [21] from Dimensional Analysis, both the overshoot and $M_S$ can be expressed in terms of the single dimensionless relation $\gamma/h$. This result has several implications:

- First, it shows that the single parameter $\gamma$ allows a quantitative tuning of the classical indicators.

- Second, it shows that it is not necessary to calculate explicitly the corresponding measures of interest (overshoot, $M_S$, etc) since the exact graphs of the corresponding (general) functions can be obtained through a single simulation experiment.

Figure 3.5 depicts the nominal overshoot and $M_S$ as functions of $\gamma/h$. In order to facilitate the usage of this information for design purposes, the following linear approximations (valid in the range $\gamma \geq 5h$) have been obtained

$$M_S = 0.06913\frac{\gamma}{h} + 1.254 \tag{3.46}$$

$$\text{Overshoot (\%)} = 5.521\frac{\gamma}{h} - 19.13 \tag{3.47}$$

It is also useful to have an idea of how uncertainty degrades performance. More concretely, next we analyze the effect of parametric uncertainty on the overshoot. Let us consider a certain amount ($\delta$) of simultaneous uncertainty on each parameter of the FOPTD model. The worst case for this uncertainty profile is given by

$$\tilde{P} = K_g(1+\delta)\frac{e^{-s(1+\delta)h}}{(1-\delta)\tau s + 1} \tag{3.48}$$

Thus, fixing $\delta$ is possible to determine the worst-case overshoot as a function of $\frac{\gamma}{h}$. Figure 3.6 does so for different values of $\delta$, whereas Table 3.1 provides simple linear approximations.

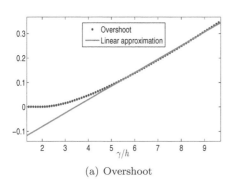 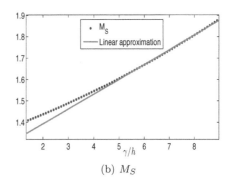

(a) Overshoot            (b) $M_S$

**FIGURE 3.5**
Overshoot and $M_S$ vs $\frac{\gamma}{h}$.

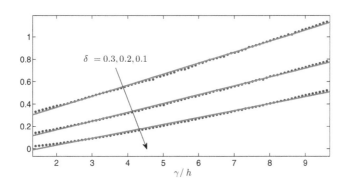

**FIGURE 3.6**
Maximum overshoots for different levels of parametric uncertainty with respect to the FOPTD model.

**Remark 3.4.1.** *It is worth noting (see Figure 3.5(b)) that robustness decreases as we get closer to the regulator mode. The cancellation (3.33) (in conjunction with the waterbed effect [72]) increases the peak on S. This can be regarded as an inherent feature of a regulator-type tuning. In Figure 3.7, this point has been exemplified.*

## 3.5    Simulation examples

This section illustrates the presented method by considering two simulation examples.

**TABLE 3.1**
Worst-case (approximate) overshoots for different levels of parametric uncertainty.

| uncertainty | worst-case overshoot |
|---|---|
| $\delta = 0.1$ | $6.2\gamma/h - 9.33$ |
| $\delta = 0.2$ | $7.86\gamma/h + 1.05$ |
| $\delta = 0.3$ | $9.86\gamma/h + 17.22$ |

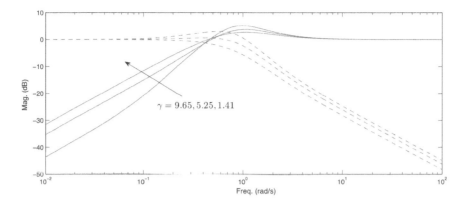

**FIGURE 3.7**
Shapes of the sensitivity (solid) and complementary (dash) sensitivity functions for $P = \frac{e^{-s}}{10s+1}$ and different tunings of $\gamma$. When $\gamma = \gamma_{ld} = 9.65$, condition (3.33) pushes up the peak of the sensitivity function.

### 3.5.1   Example 1

First, this example shows how the tuning of $\gamma$ can be done quantitatively in terms of prescribed levels for the maximum overshoot. Second, it stresses the main feature of the suggested $\gamma$-tuning procedure: smooth control with adjustable operation mode. The following model will be used

$$P = \frac{e^{-0.073s}}{1 + 1.073s} \tag{3.49}$$

It will be assumed that the real plant consists of a FOPTD process. However, 20% of parametric uncertainty with respect to the model in (3.49) will be considered. Consequently, the worst case real plant is given by

$$P_1 = \frac{1.2e^{-0.0876s}}{1 + 0.858s} \tag{3.50}$$

The following controller specifications are given:

- Obtain the best possible load disturbance attenuation ensuring that the overshoot does not exceed 50%.

From (3.34), the best load disturbance attenuation is achieved for $\gamma = 0.764$. However, this tuning would lead to a high overshoot even in the nominal response. In fact, from (3.47) the nominal overshoot is approximately 40%. On the contrary, the worst load disturbance attenuation is given by $\gamma = 0.1032$, which leads to low overshoots. The value of $\gamma$ satisfying the given requirement can be easily obtained from Table 3.1 (case $\delta = 0.2$). The exact value turns out to be $\gamma = 0.4036$. Figure 3.8 shows the time responses associated with the nominal model $P$. The responses for the case with 20% of parametric uncertainty can be seen in Figure 3.9. It is readily seen how the best possible disturbance rejection has been obtained while keeping the worst-case overshoot below the prescribed value. It should be noted that in the case of the plant not being a FOPTD process, the quantitative formulas derived in this work are not exact. In this case, the important point is that the general methodology continues to be valid as long as the plant can be adjusted well by a FOPTD model.

In [70], it is claimed that smooth control is probably the most common objective in industrial practice, where very high performance is not the main concern. Despite this, almost all published PID tuning rules aim at high performance [70]. Even in direct synthesis

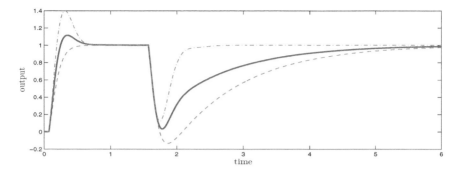

**FIGURE 3.8**
Nominal time responses for $\gamma = 0.1032$ (dash), $\gamma = 0.4036$ (solid), and $\gamma = 0.764$ (dash-dot).

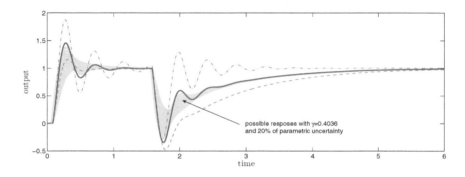

**FIGURE 3.9**
Worst-case time responses in presence of 20% of parametric uncertainty with respect to $P$ for $\gamma = 0.1032$ (dash), $\gamma = 0.4036$ (solid), and $\gamma = 0.764$ (dash-dot).

approaches like IMC, which have the closed loop time constant $\lambda$ as a tuning parameter, the emphasis is to obtain a lower bound on $\lambda$ (tight control). To clarify these points, we will consider the ISE-optimal tuning rules obtained numerically in [96] for set-point tracking and disturbance attenuation. In Figure 3.10, the responses associated with these rules and the proposed approach are compared. The tuning rules given in [96] do not specify a value for $N$ in (3.3); for the simulations, the value $N = 20$ has been used[1]. As it can be seen, the responses for the method in [96] are extremely aggressive, resulting into poor robustness. It has been found that $M_S = 2.65$ for the set-point tuning, whereas $M_S = 21.22$ (!) for the regulatory tuning, indicating that no robustness consideration was taken into account. In contrast with these negative indicators, the proposed approach yields $M_S = 1.41$ ($\gamma = T_M = 0.1032$) and $M_S = 1.98$ ($\gamma = 0.755 \approx \gamma_{ld}$), adhering to the rule of thumb $M_S < 2$ for good robustness [72]. The smoothness of the control action is also evident, and the performance level acceptable.

The IMC-based tuning in [13] has also been evaluated in the load disturbance column of Figure 3.10. Following the work in [13], we have adjusted the value of $\lambda$ to obtain tight control, i.e., the best possible disturbance attenuation with acceptable robustness ($M_S < 2$). Taking $\lambda = 0.0321$ yields $M_S = 1.98$. For smaller values of $\lambda$, the robustness margins decrease. Therefore, the $\gamma$-tuning strategy obtains considerably better regulatory performance with the same robustness level, only at the expense of a modest increment in the control effort.

---

[1]It can be seen that $N = 10$, a more common value, yields even worse results.

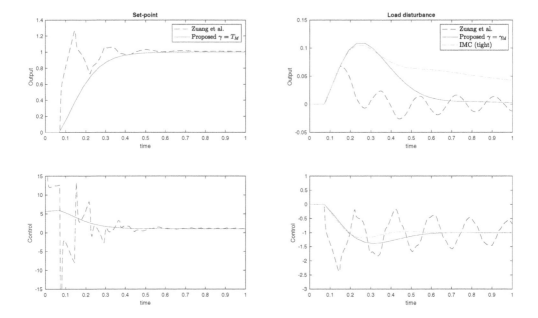

**FIGURE 3.10**
Set-point and load disturbance responses.

### 3.5.2 Example 2

The aim of this example is twofold: on the one hand, it illustrates that the proposed general methodology continues to be valid when the plant is not a FOPTD system. On the other hand, it shows that the presented approach gives good results, in terms of smooth responses, when compared with a well-known tuning rule suggested in the literature. Let us consider the isothermal series/parallel Van de Vusse reaction [79] taking place in an isothermal continuous stirred tank reactor (CSTR). The rates of formation of $A$ and $B$ are assumed to be:

$$
\begin{aligned}
r_A &= -k_1 c_A - k_3 c_A^2 \\
r_B &= k_1 c_A - k_2 c_A
\end{aligned}
$$

where $k_1 = 50 h^{-1}$, $k_2 = 100 h^{-1}$, and $k_3 = 10 \; l(\text{gmol} \cdot h)^{-1}$ are the reaction-rate constants. The feed steam consists of pure $A$. The mass balance for $A$ and $B$ is given by

$$
\begin{aligned}
V\frac{dc_A}{dt} &= F(c_{A_0} - c_A) + V(-k_1 c_A - k_3 c_A^2) \\
V\frac{dc_B}{dt} &= F(-c_B) + V(k_1 c_A - k_2 c_A)
\end{aligned}
$$

where $F$ is the inlet flow rate of product $A$, $V$ is the reactor volume which is kept constant during the operation, $c_A$ and $c_B$ are the concentrations of the species $A$ and $B$ inside the reactor, respectively, and $c_{A_0} = 10 \; \text{gmol} \cdot l^{-1}$ is the concentration of $A$ in the feed steam. We wish to maintain $c_B$ at its set-point using the dilution rate $F/V$ as the manipulated variable. Figure 3.11 depicts the CSTR control system under consideration. The system can

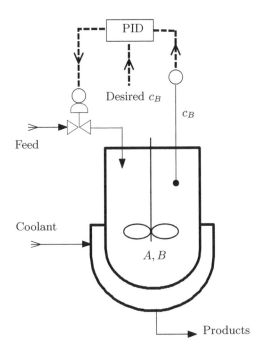

**FIGURE 3.11**
CSTR control system.

be put in the standard form

$$\dot{x}_1 = -k_1 x_1 - k_3 x_1^2 + (c_{A_0} - x_1)u$$
$$\dot{x}_2 = k_1 x_1 - k_2 x_2 - x_2 u$$
$$y = x_2$$

where $x_1 = c_A, x_2 = c_B, u = F/V, y = c_B$. Initially the system is at steady state with $x_1 = 2, x_2 = 0.851$. For control purposes, the nonlinear system is approximated at the initial steady state by a FOPTD model with $K_g = 0.033, \tau = 0.02, h = 0.005$. Figure 3.12 shows the responses of the nonlinear (real) system and the linear (FOPTD) approximation. Now we will turn our attention to the control of the described plant. For comparison purposes, we will include the well-known AMIGO tuning rule in the simulations. As it is explained in [19], the AMIGO tuning rule for the ISA PID controller was obtained to yield good disturbance attenuation along the lines of the classical Ziegler-Nichols method but with improved robustness.

In order to evaluate the performance obtained with the two methods at hand, the following standard measures will be used:

- *Output performance*: The integrated absolute error (IAE) of the error $e = r - y$ will be computed.

$$\text{IAE} = \int_0^\infty |e(t)| dt$$

- *Input performance*: To evaluate the manipulated input usage, the total variation (TV) of the control signal $u(t)$ will be computed.

$$\text{TV} = \int_0^\infty |\dot{u}(t)| dt$$

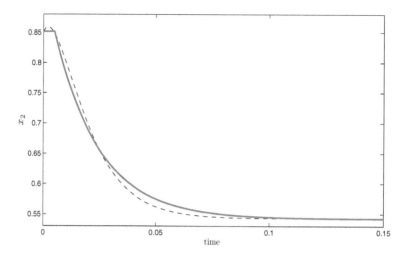

**FIGURE 3.12**
Real nonlinear system (dashed) and FOPTD (solid) approximation.

In addition, the percent overshoot of the output $y(t)$, denoted by $y_{ov}$, will be taken into account for set-point output performance. The results obtained are depicted in Figure 3.13 and summarized in Table 3.2.

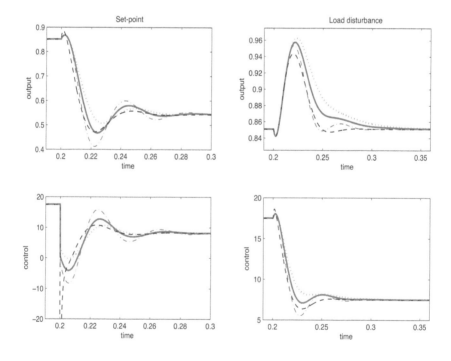

**FIGURE 3.13**
Time responses for the proposed: $\gamma = 0.0071 = T_M$ (dot), $\gamma = 0.032 = \gamma_{ld}$ (dash-dot), $\gamma = 0.0195$ (bold solid), and AMIGO (dashed) tuning rules.

**TABLE 3.2**

Results for set-point and load disturbance.

| Tuning method | SP | | | LD | |
|---|---|---|---|---|---|
| | IAE | $y_{ov}$ | TV | IAE | TV |
| Proposed ($\gamma = 0.0071$) | 0.048 | 7.02% | 32.28 | 0.0044 | 11 |
| Proposed ($\gamma = 0.032$) | 0.060 | 31.85% | 66.8 | 0.0024 | 16.96 |
| Proposed ($\gamma = 0.0195$) | 0.0051 | 16.23% | 46.7 | 0.0034 | 13.03 |
| AMIGO | 0.0041 | 15.15% | 158.86 | 0.0023 | 14.7 |

It can be seen that the AMIGO tuning provides better results in general. The reason for this superior performance relies on the numerical, non-convex optimization approach on which it is based. However, it requires a more aggressive control effort in the set-point response. Overall, it can be observed that the proposed tuning yields smooth control, while allowing for an easy adjustment of the operation mode of the controller.

# 4

## Analysis of the Smooth/Tight–Servo/Regulation Tuning Approaches

Chapter 2 presented a simple (IMC-like) analytical design based on a WSP. The tuning of the controller involved a single tuning parameter ($\lambda$), closely related to the closed-loop bandwidth. It was shown that, for lag-dominant processes, the disturbance attenuation was sluggish. On the other hand, Chapter 3 revised a model matching strategy for improving the load disturbance response. More concretely, a tuning parameter ($\gamma$) was used to account for a servo/regulation balance. This chapter compares the $\lambda$ and $\gamma$-tuning methods from a servo/regulation point of view. Simulation examples clarify the discussion and confirm the effectiveness (and the limits) of each approach.

The organization of this chapter is as follows. First of all, a short review of the designs obtained so far is presented. Next, Section 4.2 proposes a tuning interval for $\lambda$, and suggests a value for balanced smooth/tight control. Section 4.3 is devoted to compare, from a servo/regulation point of view, the $\lambda$ and $\gamma$-tuning approaches. Section 4.4 deals with some implementation aspects, whereas simulation examples are given in Section 4.5

## 4.1 Revisiting the model-matching designs

In Chapter 3, the following controller was derived:

$$K = \frac{1}{K_g(\rho + \gamma)} \frac{(1 + \tau s)(1 + \chi s)}{s(1 + \frac{zT_M + h\chi}{\rho + \gamma}s)} \tag{4.1}$$

where $K_g, \tau, h$ constitute the FOPTD model information; $\chi = z + h - \rho + T_M - \gamma$ and $z, T_M$ were fixed in (3.25). It was shown in Section 3.4 that if one considers disturbances entering at the output of the plant, choosing $\gamma = \gamma_o = 4.56h$ provides the best disturbance attenuation with respect to the ISE criterion. In addition, it was shown that if the disturbance occurs at the input of the plant, the disturbance rejection produced by taking $\gamma = \gamma_o = 4.56h$ may still be improved significantly. At the end, the following interval for $\gamma$ was found

$$\gamma \in [\gamma_{sp} = \sqrt{2}h, \gamma_{ld}] \tag{4.2}$$

where

$$\gamma_{ld} \approx -\frac{12.36h(-\tau + \sqrt{2}h)}{h + \tau} \tag{4.3}$$

and extreme values for $\gamma$ represent the tuning for *servo* ($\gamma = \gamma_{sp}$) and *regulator* ($\gamma = \gamma_{ld}$) operation. In general, the greater the value of $\gamma$, the greater the overshoot in the set-point response and the smaller the stability margins.

From the point of view of balanced servo/regulation operation, this chapter compares the control settings discussed in Chapter 3 [5] through the $\gamma$-tuning procedure with those obtained using the simpler IMC-like design of Chapter 2 [8]. Within the latter method, the tuning parameter $\lambda$ provides a means to adjust the *robustness/performance* trade-off. In this situation, we can distinguish between two tuning strategies [70, 13]:

- *Smooth* control, which corresponds to the slowest-possible control with acceptable disturbance rejection. In Chapter 2 [8], the value $\lambda = h$ was selected for smooth control, yielding $M_S \approx 1.42$.

- *Tight* control, which obeys the fastest-possible control with acceptable robustness. This option implies a reduction in the value of $\lambda$, which can be used to improve the regulatory performance.

How to tune $\lambda$ for tight control is addressed in this chapter. Afterwards, a compromise between smooth and tight control is determined, corresponding to an intermediate value for $\lambda$.

## 4.2   Smooth/tight tuning

Based on a FOPTD model, the controller obtained in Chapter 2 is given by

$$K = \frac{1}{K_g} \frac{(\frac{h}{2}s + 1)(\tau s + 1)}{s(\lambda^2 s + 2\lambda + \frac{h}{2})} \tag{4.4}$$

and depends on a single tuning parameter: $\lambda$. As in the standard IMC procedure [49], the role of $\lambda$ is to provide the necessary roll-off at high frequencies. First, let us consider the following question: How to choose $\lambda$ for optimal load disturbance? We know that the optimal behaviour is recovered for $\lambda \to 0$. However, this would imply no robustness according to the detuning role of $\lambda$. It is easy to see[1] that the value $\lambda \approx 0.22h$ provides the best disturbance rejection with respect to disturbances entering at the output of the plant and the ISE criterion. However, the corresponding robustness for $\lambda \approx 0.22h$ is poor. Augmenting $\lambda$ increases the value of the ISE also for disturbances entering at the input. Consequently, according to the *tight* control concept, the idea is to select the smallest possible value (and consequently, the best possible load disturbance rejection) for $\lambda$ providing acceptable robustness. Stated otherwise, since the IMC-like approach cannot produce the cancellation in (3.33), the best that can be done to improve the load disturbance attenuation is making the closed-loop system faster. Sensitivity to modelling errors can be captured by the peak of the sensitivity function:

$$M_S \doteq \|S(j\omega)\|_\infty \doteq \max_\omega \left| \frac{1}{1 + L(j\omega)} \right| \tag{4.5}$$

By applying the Buckingham pi theorem [21] from dimensional analysis, $M_S$ can be determined in terms of a unique dimensionless parameter: $\lambda/h$. In concrete, $\lambda = 0.56h$ provides $M_S \approx 1.75$. We will consider this value of $\lambda$ for *tight* control.

On the other hand, we saw in Chapter 2 that by choosing $\lambda = h$, point on which $M_S \approx 1.42$, smooth set-point responses are obtained. The conclusion of the above analysis is that the interval to be considered for the *smooth/tight* trade-off is

$$\lambda = [\lambda_{ld} = 0.56h, \lambda_{sp} = h] \tag{4.6}$$

---

[1]Consult the Appendix A for the details.

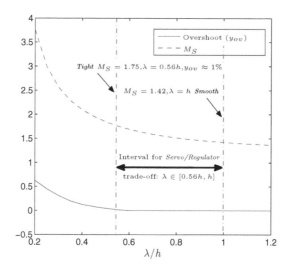

**FIGURE 4.1**
Sensitivity peak ($M_S$) and output overshoot against $\lambda/h$. Smooth and tight control mode tuning.

where the extreme values represent the *tight* ($\lambda = 0.56h$) and *smooth* ($\lambda = h$) tunings for the proposed controller.

Now, we will inspect when the $\lambda$-tuning approach is suitable to provide balanced servo/regulation operation, comparing it with the $\gamma$-tuning procedure. The first issue is how to adjust the value of $\lambda$, as Figure 4.1 illustrates.

Let us define

$$\Delta(ISE)(\lambda/h) = \frac{ISE(\lambda/h, 1) - ISE(0.56, 1)}{ISE(0.56, 1)} \tag{4.7}$$

and

$$\Delta(M_S)(\lambda/h) = \frac{M_S(\lambda/h) - 1.42}{1.42} \tag{4.8}$$

as measures of the relative load disturbance performance and robustness degradation with respect to the corresponding optimal values. The ISE degradation index captures how the ISE caused by a disturbance degrades with $\lambda/h$. In the case of load disturbance peformance, the minimum value is obtained for $\lambda = 0.56h$. As for the minimum value of $M_S$, it corresponds to 1.42. Now, it stands to reason to determine the trade-off tuning $\lambda$ value as

$$(\lambda/h)_{\text{trade-off}} = \arg\min_{\lambda/h} \max \left\{ \Delta(ISE), \Delta(M_S) \right\} \tag{4.9}$$

Figure 4.2 displays the situation graphically. The trade-off value for $\lambda$ is finally found to be $\lambda = 0.7h$. This value can be taken as a good rule of thumb as it will be seen later on.

In the next subsection, we explore in more depth the applicability of the suggested $\lambda$-tuning, comparing it with the $\gamma$-tuning strategy of Chapter 3.

## 4.3 Servo/regulation tuning

The $\lambda$-tuning strategy yields good set-point and disturbance responses when disturbances enter at the output of the plant. However, for lag-dominant plants, it provides

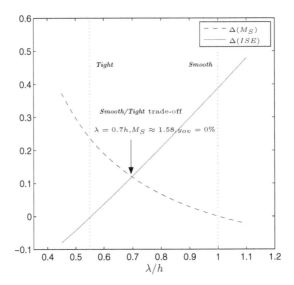

**FIGURE 4.2**
Smooth/tight trade-off tuning. Normalized $M_S$ and $ISE$ degradations against $\lambda/h$.

sluggish (load) disturbance response. This way, in this comparison, we will focus on load disturbance rejection capabilities. Let us define the load disturbance performance degradation with respect to the $\gamma$-tuning method as follows

$$\Delta_{LD} = \frac{ISE(LD)_{\lambda-tuning} - ISE(LD)_{\gamma-tuning}}{ISE(LD)_{\gamma-tuning}} \qquad (4.10)$$

Based on dimensional analysis [21], it can be seen that $\Delta_{LD}$ only depends on $\frac{h}{\tau}$ once the values for $\lambda$ and $\gamma$ have been fixed. Remember that we are interested in obtaining a compromise between the servo and the regulator modes. In order to establish a comparison framework, let us assume that the maximum nominal overshoot is 10%. For a lag-dominant plant, it can be seen that this specification is met for $\gamma \leq 5.25h$. Figure 4.3 displays $\Delta_{LD}$

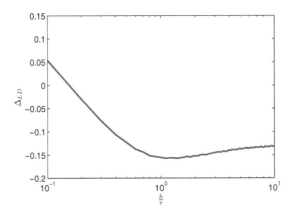

**FIGURE 4.3**
Load disturbance performance degradation of the $\lambda$-tuning method ($\lambda = 0.7h$) with respect to the $\gamma$-tuning one ($\gamma = 5.25h$) in *servo/regulation* trade-off mode.

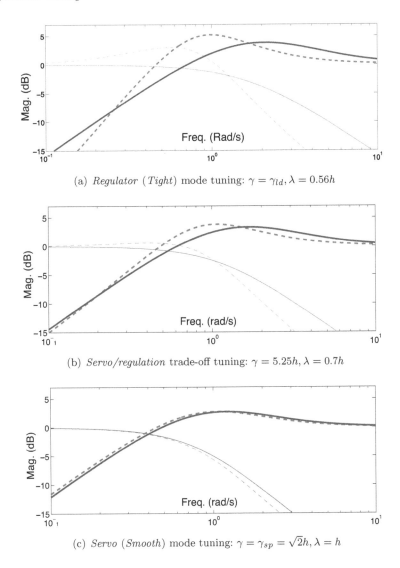

(a) *Regulator (Tight)* mode tuning: $\gamma = \gamma_{ld}, \lambda = 0.56h$

(b) *Servo/regulation* trade-off tuning: $\gamma = 5.25h, \lambda = 0.7h$

(c) *Servo (Smooth)* mode tuning: $\gamma = \gamma_{sp} = \sqrt{2}h, \lambda = h$

**FIGURE 4.4**
Sensitivity (thick) and complementary sensitivity (thin) functions for the $\gamma$-tuning (dashed) and $\lambda$-tuning (solid) approaches in *Regulator (Tight)*, *Servo (Smooth)* and trade-off modes for the plant $P = \frac{-0.5s+1}{(0.5s+1)(10s+1)} \approx \frac{e^{-s}}{10s+1}$.

for $\gamma = 5.25h$ and $\lambda = 0.7h$. It can be observed that the $\lambda$-tuning provides better disturbance rejection than the $\gamma$-tuning design in the balanced lead/lag and lead-dominant cases. However, for $\frac{h}{\tau} < 0.175$ the $\gamma$-tuning method provides better disturbance attenuation.

In order to outline the basic differences between the two approaches, let us take a look at the (approximate) sensitivity and complementary sensitivity functions depicted in Figure 4.4. The $\gamma$-tuning and the $\lambda$-tuning methodologies provide almost the same design when tuned in servo (smooth) mode; see Figure 4.4(c). As we move towards the regulator (tight) mode, it is clear that $M_S$ increases for the $\gamma$-tuning approach. As it was discussed in Chapter 3, the reason for this greater peak can be attributed to condition (3.33).

## 4.4   Implementation aspects

So far in this chapter, the controller has been taken in the second-order form

$$K = \frac{c_1 s^2 + c_2 s + c_3}{s(d_1 s + 1)} \tag{4.11}$$

where $c_1, c_2, c_3, d_1$ are assumed to be positive real constants. The controller above has four degrees of freedom and can be considered a *general* PID controller. However, for implementation purposes, it is desirable to cast this general form into practical PID realizations. In Chapters 2 and 3, the ISA PID structure [20] was considered:

$$K = K_c \left( 1 + \frac{1}{sT_i} + \frac{sT_d}{1 + s\frac{T_d}{N}} \right) \tag{4.12}$$

The controller above has the three modes ($K_c, T_i, T_d$) working additively, and it is sometimes referred to as *noninteractive, ideal,* or *parallel.* The derivative filter parameter $N$ is typically fixed by the manufacturer, being $N = 10$ a typical value [51]. However, as reported in [36], fixing $N$ independently of the other three parameters may not be a good idea. This can be observed by posing (4.12) in the general second-order form (4.11)

$$K = K_c \frac{T_d \left(1 + \frac{1}{N}\right) s^2 + \left(1 + \frac{T_d}{T_i N}\right) s + \frac{1}{T_i}}{s \left(\frac{T_d}{N} s + 1\right)} \tag{4.13}$$

It is clear that the filter derivative parameter $N$ has a big influence on the coefficients of the second-order controller. Consequently, PID design should be a four-parameter design including $N$ [36] (this was also the case in Chapters 2 and 3). As it is claimed in [46], the use of the fixed derivative filter parameter $N$ explains in part the industrial myth that *derivative action does not work.* Another problem with the ISA PID form can be detected if we convert from (4.11) to (4.12):

$$
\begin{aligned}
K_c &= c_2 - c_3 d_1 \\
T_d &= \frac{c_1}{K_c} - d_1 \\
T_i &= \frac{K_c}{c_3} \\
N &= \frac{T_d}{d_1}
\end{aligned}
\tag{4.14}
$$

It is evident that there are combinations of $c$ and $d$ that cannot be described. For example, the tuning rule for the proposed controller with $\lambda = 0.7h$ is indicated in Table 4.1.

It is easy to see that both $N$ and $T_d$ may become negative, which is physically unfeasible and restricts the application of the proposed method. As pointed out in [46] regarding

**TABLE 4.1**
Tuning rule for the proposed controller assuming the ISA PID algorithm.

| $K_c$ | $T_i$ | $T_d/N$ | $N+1$ |
|-------|-------|---------|-------|
| $\frac{0.53}{k_g}\frac{T_i}{h}$ | $\tau + 0.25h$ | $0.258h$ | $1.94\frac{\tau}{T_i}$ |

**TABLE 4.2**
Tuning rule for the controller (4.4) in the ideal, output-filtered form ($\lambda = 0.7h$).

| $\mathbf{K_c}$ | $\mathbf{T_i}$ | $\mathbf{T_d}$ | $\mathbf{T_F}$ |
|---|---|---|---|
| $\frac{1}{k_g}\frac{T_i}{1.9h}$ | $\tau + \frac{h}{2}$ | $\frac{\tau h}{2T_i}$ | $0.2579h$ |

model-based designs, the following practical form introduced in [49] is preferable

$$K = K_c \left(1 + \frac{1}{T_i s} + T_d s\right)\frac{1}{T_F s + 1} \tag{4.15}$$

This alternative form provides a more straightforward parameterization of the second-order controller (4.11):

$$\begin{aligned} K_c &= c_2 \\ T_d &= \frac{c_1}{K_c} \\ T_i &= \frac{K_c}{c_3} \\ T_F &= d_1 \end{aligned} \tag{4.16}$$

The corresponding tuning rule is given in Table 4.2.

Note that it undergoes the limitations found in the ISA PID case. The advantages of using the four-parameter, parallel, output-filtered PID structure (4.15) in conjunction with the IMC method can be consulted in [46]. Due to all these considerations, the PID form (4.15) is the recommended one for the presented designs.

## 4.5   Simulation examples

The purpose of this section is to compare by example the $\lambda$ and $\gamma$-tuning methods. The experimental setup is summarized in Table 4.3, and it has been taken from [19] (Examples 1, 3, 4) and [16] (Example 2).

As it can be seen, we will adopt a FOPTD model for four different plants. Simulations will be shown for the nominal case—i.e., we will assume that the real plant corresponds

**TABLE 4.3**
Experimental setup.

| Example | Model | $\frac{h}{\tau}$ | Real plant CASE I | CASE II |
|---|---|---|---|---|
| 1 | $\frac{e^{-s}}{1+0.093s}$ | 10.75 | $\frac{e^{-s}}{(1+0.05s)^2}$ | $\frac{1.2e^{-1.2s}}{1+0.0744s}$ |
| 2 | $\frac{e^{-0.99s}}{1+1.65s}$ | 0.6 | $\frac{e^{-0.5s}}{(s+1)^2}$ | $\frac{1.2e^{-1.188s}}{1+1.32s}$ |
| 3 | $\frac{e^{-1.42s}}{1+2.9s}$ | 0.49 | $\frac{1}{(s+1)^4}$ | $\frac{1.2e^{-1.704s}}{1+2.32s}$ |
| 4 | $\frac{e^{-0.073s}}{1+1.073s}$ | 0.068 | $\frac{1}{(1+s)(1+0.1s)(1+0.01s)(1+0.001s)}$ | $\frac{1.2e^{-0.0876s}}{1+0.8584s}$ |

exactly to the model—as well as for the uncertain scenario. Two cases will be distinguished here:

- CASE I: The real plant will not be a FOPTD model. This case will tackle the usual situation in which a simple model is used to represent a real (more complex) system, giving rise to neglected/unmodeled dynamics.

- CASE II: The real plant will be assumed to be perfectly described by the model. But 20% of parametric uncertainty will be taken into account.

The tuning of $\gamma$ for the design of Chapter 3 will be done so as to get the best possible results limiting the nominal overshoot to a 10% value. This performance specification obeys the fact that in many processes an excessive overshoot is not acceptable. Figures 4.5, 4.7, 4.9 concern the lead-dominant and balanced lag and delay cases (Examples 1–3) and it is

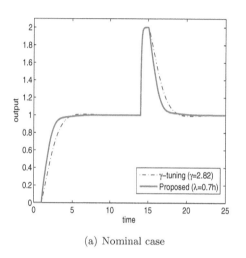

(a) Nominal case

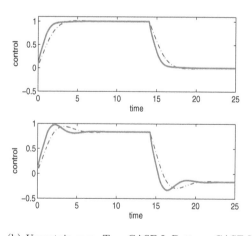

(b) Uncertain case. Top: CASE I. Bottom: CASE II

**FIGURE 4.5**
Time responses for Example 1.

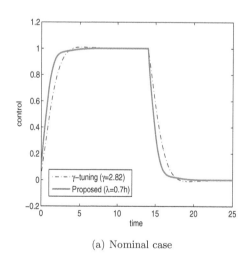

(a) Nominal case

(b) Uncertain case. Top: CASE I. Bottom: CASE II

**FIGURE 4.6**
Control effort for Example 1.

shown, especially for the lead-dominant plant (Example 1), that the $\lambda$-tuning is superior to the $\gamma$-tuning strategy, achieving a better trade-off tuning.

The most interesting case is the *lag-dominant* one, which is dealt with in Example 4. The corresponding simulation results are captured in Figure 4.11. In this case, a trade-off tuning is more difficult to achieve. The $\gamma$-tuning produces excellent load disturbance attenuation by selecting $\gamma = \gamma_{ld} = 0.76$. Nevertheless, this choice produces an excessive nominal overshoot (40%). The value of $\gamma$ needs to be decreased until $\gamma = 0.38$ in order to reduce the overshoot to approximately 10%. When this is done, the load disturbance attenuation gets approximately the same as that obtained with the IMC-like approach. Regarding the set-point response, the suggested method provides clear superiority.

The results obtained for the Examples 1–4 are collected and quantified in Tables 4.4 and 4.5 for the sake of a more precise comparison. Figures 4.6, 4.8, 4.10, 4.12 plot the required

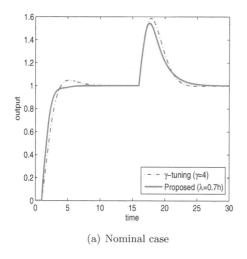

(a) Nominal case

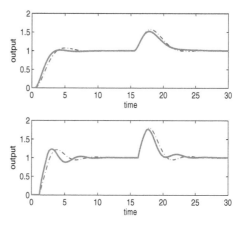

(b) Uncertain case. Top: CASE I. Bottom: CASE II

**FIGURE 4.7**
Time responses for Example 2.

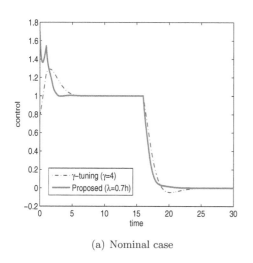

(a) Nominal case

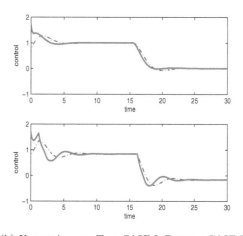

(b) Uncertain case. Top: CASE I. Bottom: CASE II

**FIGURE 4.8**
Control effort for Example 2.

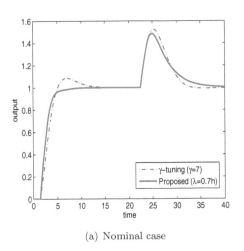 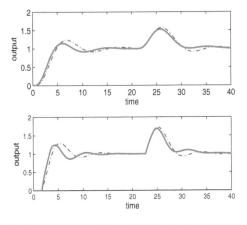

(a) Nominal case          (b) Uncertain case. Top: CASE I. Bottom: CASE II

**FIGURE 4.9**
Time responses for Example 3.

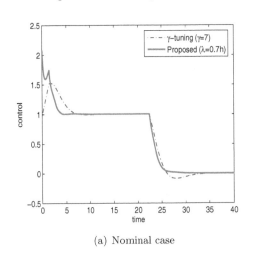 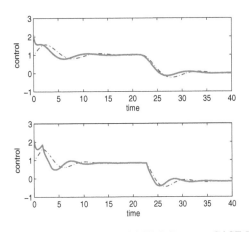

(a) Nominal case          (b) Uncertain case. Top: CASE I. Bottom: CASE II

**FIGURE 4.10**
Control effort for Example 3.

**TABLE 4.4**
Performance evaluation for the nominal case.

| Example | Tuning | $SP_{ISE}$ | $LD_{ISE}$ | $M_S$ |
|---|---|---|---|---|
| 1 | $\lambda$-tuning $(\lambda = 0.7)$ | 1.48 | 1.44 | 1.58 |
|   | $\gamma$-tuning $(\gamma = 2.82)$ | 1.82 | 1.76 | 1.48 |
|   |   |   |   |   |
| 2 | $\lambda$-tuning $(\lambda = 0.69)$ | 1.47 | 0.67 | 1.58 |
|   | $\gamma$-tuning $(\gamma = 4)$ | 1.72 | 0.85 | 1.55 |
|   |   |   |   |   |
| 3 | $\lambda$-tuning $(\lambda = 0.99)$ | 2.1 | 0.85 | 1.58 |
|   | $\gamma$-tuning $(\gamma = 7)$ | 2.43 | 1 | 1.6 |
|   |   |   |   |   |
| 4 | $\lambda$-tuning $(\lambda = 0.05)$ | 0.1085 | 0.0084 | 1.58 |
|   | $\gamma$-tuning $(\gamma = 0.38)$ | 0.1243 | 0.0077 | 1.62 |
|   | $\gamma$-tuning $(\gamma = 0.76)$ | 0.139 | 0.0026 | 1.99 |

**TABLE 4.5**
Performance evaluation for the uncertain case.

| Example | Tuning | CASE I | | | CASE II | | |
|---|---|---|---|---|---|---|---|
| | | $\frac{SP}{ISE}$ | $y_{ov}$ (%) | $\frac{LD}{ISE}$ | $\frac{SP}{ISE}$ | $y_{ov}$ (%) | $\frac{LD}{ISE}$ |
| 1 | $\lambda$-tuning | 1.49 | 0 | 1.45 | 1.59 | 16.8 | 2.23 |
| | $\gamma$-tuning ($\gamma$=2.82) | 1.82 | 1 | 1.78 | 1.86 | 12.7 | 2.63 |
| 2 | $\lambda$-tuning | 1.42 | 1.66 | 0.66 | 1.55 | 22.68 | 0.95 |
| | $\gamma$-tuning ($\gamma$=4) | 1.68 | 7.15 | 0.85 | 1.75 | 21.24 | 1.22 |
| 3 | $\lambda$-tuning | 2.25 | 14.2 | 0.95 | 2.24 | 24.36 | 1.16 |
| | $\gamma$-tuning ($\gamma$=7) | 2.66 | 23.2 | 1.17 | 2.56 | 29.2 | 1.43 |
| 4 | $\lambda$-tuning | 0.11 | 8.7 | 0.0085 | 0.1212 | 35 | 0.009 |
| | $\gamma$-tuning ($\gamma$=0.38) | 0.13 | 18.6 | 0.0078 | 0.144 | 43 | 0.009 |
| | $\gamma$-tuning ($\gamma$=0.76) | 0.15 | 46 | 0.0027 | 0.2366 | 86 | 0.0042 |

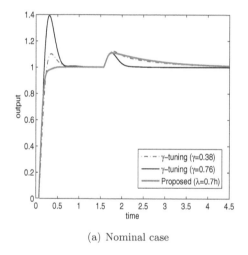

(a) Nominal case

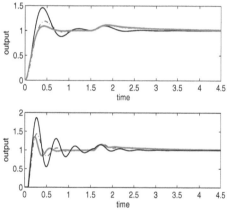

(b) Uncertain case. Top: CASE I. Bottom: CASE II

**FIGURE 4.11**
Time responses for Example 4.

control effort for the two methods at hand. It is noticeable that the $\gamma$-tuning results in less a demanding controller generally. In spite of this, the manipulated variable movements are still quite acceptable for the IMC-like design tuned with $\lambda = 0.7h$. If strictly necessary, the control action *kicks* can be reduced by slightly augmenting the value of $T_F$ in the tuning rule of Table 4.2. This is illustrated for Example 4. In this case, it can be seen that $T_F = 0.0188$. Figures 4.13 and 4.14 show that by increasing this value—corresponding to the $d_1$ coefficient with respect to (4.11)—the required control effort can be made equal to that of the $\gamma$-tuning method while still producing quite similar set-point and load disturbance responses. It has been verified that that $M_S = 1.618$ for $T_F = 0.0288$, whereas $M_S = 1.64$ for $T_F = 0.0388$.

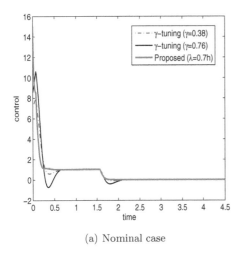

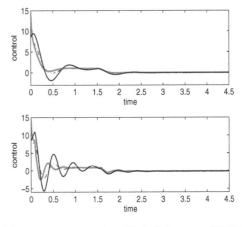

(a) Nominal case        (b) Uncertain case. Top: CASE I. Bottom: CASE II

**FIGURE 4.12**
Control effort for Example 4.

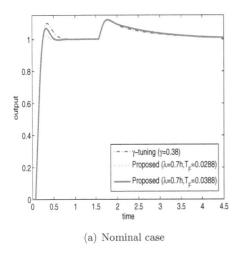

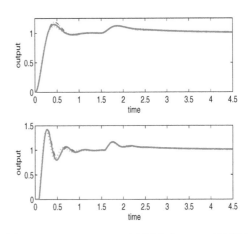

(a) Nominal case        (b) Uncertain case. Top: CASE I. Bottom: CASE II

**FIGURE 4.13**
Time responses for Example 4 with increased value of $T_F$.

## 4.6   Summary

This chapter has compared the $\lambda$- and $\gamma$-tuning approaches (Chapters 2 and 3). First, the rule $\lambda = 0.7h$ has been proposed for obtaining balanced performance between the servo and regulator modes based on a trade-off between smooth and tight control. The servo/regulation trade-off limits within the $\lambda$-tuning method have been explored; if the allowed nominal overshoot is required to be inferior to a 10%, the $\lambda$-tuning method provides in general both better servo and regulatory responses for plants such that $h/\tau \geq 0.1$. Nevertheless, for larger allowed overshoots and/or more lag-dominant plants, then the $\gamma$-tuning technique allows better (input) disturbance attenuation for a given degree of robustness.

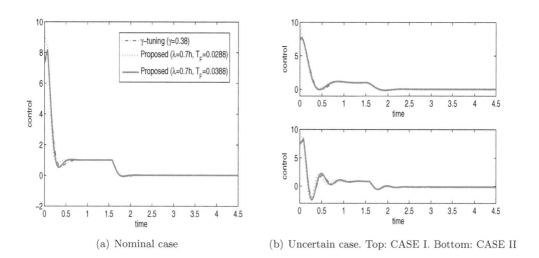

(a) Nominal case        (b) Uncertain case. Top: CASE I. Bottom: CASE II

**FIGURE 4.14**
Control effort for Example 4 with increased value of $T_F$.

# Part II

# Weight Selection for Sensitivity Shaping

# 5

# $\mathcal{H}_\infty$ Design with Application to PI Tuning

This chapter presents an $\mathcal{H}_\infty$ design that alleviates some difficulties with standard IMC, while still obeying the same spirit of simplicity. The controller derivation is carried out analytically based on a *weighted sensitivity* formulation. The corresponding frequency weight, chosen systematically, involves two tuning parameters with clear meaning in terms of common design specifications: one adjusts the *robustness/performance* trade-off as in the IMC procedure; the other one balances the *servo* and *regulatory* performance. For illustration purposes, the method is applied to analytical tuning of PI compensators. Due to its simplicity and effectiveness, the presented methodology is also suitable for teaching purposes.

The rest of the chapter is organised as follows: Section 5.2 presents the analytical solution to the proposed design method, based on the $\mathcal{H}_\infty$ WSP, a generic solution, independent of the specific weight. This is followed in Section 5.3 with an analysis of the solution in terms of a generic weighting function but with a specific and motivated structure. Section 5.4 introduces some stability and robustness characteristics in terms of the dual-locus diagramme, whereas Section 5.5 deals with its application to analytical tuning of PI controllers. Simulation examples are given in Section 5.6 to emphasize the new features of the proposed approach.

## 5.1 Problem statement

Simplicity is a desired feature of a control algorithm: we would like it to be widely applicable and easy to understand, involving as few tuning parameters as possible. Ideally, these parameters should possess a clear engineering meaning, making the tuning a systematic task according to the given specifications. As for implementation, low-order controllers are preferable.

In this line, the PID controller is recognized to be the bread and butter of automatic control, being by far the most dominating form of feedback in a wide range of industrial applications; the PID strategy is particularly effective in process control, where a combination of benign process dynamics and modest performance requirements finds its place [20]. The ideal PID law is based on the present (P), past (I), and estimated future (D) error information. In accordance with this original conception, there are only three tuning parameters. Even for such a simple strategy, it is not easy to find good settings without a systematic procedure [57, 69, 52].

During the last twenty years, there has been a revived interest in PID control, motivated by the advent of model predictive control, which requires well-tuned PID compensators at the bottom level, and the emergence of auto-tuning tools [18]. As a result, numerical (optimization-based) techniques have been suggested in the literature [96, 86, 19, 78]. In the same vein, analytically-derived tuning rules have appeared [32, 43, 64, 81]. Another reason for the PID revival has been the lack of results regarding stabilizing delayed systems [68, 33, 74, 53]. These research efforts, especially the trend for analytical design, has incorporated into the PID arena the control theory mainstream developments, leaving aside more specific techniques.

Among the analytical methods, IMC [49] has gained remarkable industrial acceptance due to its simple yet effective procedure [69, 25]. IMC theory was first applied to PID control of stable plants in [61], solving the robustness problems associated with some early tunings like [97]. Although the IMC-PID settings [61] are robust and yield good set-point responses, they result in poor load disturbance rejection for integrating/lag-dominant plants [23, 34]. Alternative PID tuning rules aimed at good regulatory performance can be consulted in [34, 64]. In [69], remarkably simple tuning rules, which provide balanced servo/regulation performance, are proposed based on a modification of the settings in [61]. It is important to realize that the problems with the original IMC-based tunings come indeed from inherent shortcomings of the IMC procedure, thoroughly revised in [25].

The purpose of this chapter is to present an $\mathcal{H}_\infty$ design, which avoids some of the limitations of the IMC method, while retaining its simplicity as much as possible. In particular, the method is devised to work well for plants of modest complexity, for which analytical PID tuning is plausible.

Roughly speaking, the design procedure associated with modern $\mathcal{H}_\infty$ control theory involves the selection of frequency weights, which are used to shape prescribed closed-loop transfer functions. Many practitioners are reluctant to use this methodology because it is generally difficult to design the frequency weights properly. At the end of the day, it is quite typical to obtain high-order controllers, which may require the use of model-order reduction techniques. Apart from the cumbersome design procedure, control engineers usually find the general theory difficult to master as well. To alleviate the above difficulties, we rely here on the plain $\mathcal{H}_\infty$ WSP. By investigating its analytical solution, the involved frequency weight is chosen systematically in such a way that a good design in terms of basic conflicting trade-offs can be attained. The main contributions of the proposed procedure are:

1. The selection of the weight is *systematic* (this is not common in $\mathcal{H}_\infty$ control) and *simple*, only depending on two types of parameters:

    - One adjusts the robustness/performance trade-off as in the IMC approach.
    - The other one shifts the performance between the servo and regulator modes. As it will be explained, this can be interpreted in terms of a mixed $S/SP$ sensitivity design.

2. The method is *general*: both stable and unstable plants are dealt with in the same way. This differs from other analytical $\mathcal{H}_\infty$ procedures.

3. The controller is derived *analytically*. For simple models, this leads to well-motivated PID tuning rules which consider the stable/unstable plant cases simultaneously.

## 5.2   Analytical solution

The proposed approach stems from considering the WSP [90, 72]:

$$
\begin{aligned}
|\rho| &= \min_{K \in \mathcal{C}} \|\mathcal{N}\|_\infty \\
&= \min_{K \in \mathcal{C}} \left\| \mathcal{F}_l \left( \left[ \begin{array}{c|c} W & -WP \\ \hline 1 & -P \end{array} \right], K \right) \right\|_\infty \\
&= \min_{K \in \mathcal{C}} \|WS\|_\infty
\end{aligned}
\tag{5.1}
$$

The problem stated here is more general as it considers the possibility of the unstable cases for both $P$ and $W$. Before selecting $W$ to shape $S$, we will look for an analytical solution of (5.1). The classical design found in [28, 26] consists of transforming (5.1) into a MMP using the Youla-Kucera parameterization [89]. From an analytical point of view, the problem with this parameterization is the need of computing a coprime factorization when $P$ is unstable. In order to deal with stable and unstable plants in a unified way, it would be desirable to avoid any notion of coprime factorization. Towards this objective, the key point is to use a possibly unstable weight.

**Theorem 5.2.1.** *Assume that $P$ is purely rational (i.e., there is no time delay in $P$) and has at least one right half-plane (RHP) zero. Take $W$ as a MP weight including the unstable poles of $P$. Then, the optimal weighted sensitivity in problem (5.1) is given by*

$$\mathcal{N}^o = \rho \frac{q(-s)}{q(s)} \tag{5.2}$$

*where $\rho$ and $q = 1 + q_1 s + \cdots + q_{\nu-1} s^{\nu-1}$ (Hurwitz) are uniquely determined by the interpolation constraints:*

$$W(z_i) = \mathcal{N}^o(z_i) \qquad i = 1 \dots \nu, \tag{5.3}$$

*being $z_1 \dots z_\nu$ ($\nu \geq 1$) the RHP zeros of $P$.*

*Proof.* The following *change of variable* (or IMC parameterization [49])

$$K = \frac{Q}{1 - PQ} \tag{5.4}$$

puts $H(P, K)$ in the simpler form

$$H(P, K) = \begin{bmatrix} PQ & (1 - PQ)P \\ Q & 1 - PQ \end{bmatrix} \tag{5.5}$$

As shown in [49], internal stability is then equivalent to

- $Q \in \mathcal{RH}_\infty$

- $S = 1 - PQ$ has zeros at the unstable poles of $P$

The weighted sensitivity $WS = W(1 - PQ) = \mathcal{N}^o$ in (5.2) is achieved by

$$Q_0 = P^{-1}(1 - \mathcal{N}^o W^{-1}) \tag{5.6}$$

First, we must verify that $Q_0$ is internally stabilizing. That $Q_0 \in \mathcal{RH}_\infty$ follows from the interpolation constraints (5.3). On the other hand, $S = 1 - PQ_0 = \mathcal{N}^o W^{-1}$ is such that $S = 0$ at the unstable poles of $P$ (because $W$ contains them by assumption). Now that internal stability has been verified, it remains to be proved that $Q_0$ (equivalently $\mathcal{N}^o$) is optimal. For this purpose, we use the result, proved in [49], that the set of internally stabilizing $Q$'s can be expressed as

$$\mathcal{Q} = \{Q : Q = Q_0 + \Upsilon Q_1\} \tag{5.7}$$

where $Q_1 \in \mathcal{RH}_\infty$ is any stable transfer function, and $\Upsilon \in \mathcal{RH}_\infty$ has (exclusively) two zeros at each closed RHP pole of $P$ (the exact shape of $\Upsilon$ is not necessary for the proof). Hence, any admissible weighted sensitivity has the form

$$\begin{aligned} W(1 - PQ) &= W(1 - P[Q_0 + \Upsilon Q_1]) \\ &= W(1 - PQ_0) - WP\Upsilon Q_1 \\ &= \mathcal{N}^o - WP\Upsilon Q_1 \end{aligned}$$

Minimizing $\|\mathcal{N}^o - WP\Upsilon Q_1\|_\infty$ is a standard MMP in terms of $Q_1$, with $T_1 = \mathcal{N}^o \in \mathcal{RH}_\infty$, $T_2 = WP\Upsilon \in \mathcal{RH}_\infty$. By Lemma 2.1.1, we know that the optimal error $\mathcal{E}^o = T_1 - T_2Q_1$ is all-pass and completely determined by the RHP zeros of $T_2$, which are those of $P$. More concretely, for each RHP zero of $P$, we have the interpolation constraint $\mathcal{E}^o(z_i) = \mathcal{N}^o(z_i)$. Obviously, this implies that $\mathcal{E}^o = \mathcal{N}^o$. Equivalently, the optimal solution is achieved for $Q_1 = 0$, showing that $Q_0$ is indeed optimal. $\square$

Once the optimal weighted sensitivity has been determined, the following corollary of Theorem 5.2.1 gives the corresponding (complementary) sensitivity function and feedback controller:

**Corollary 5.2.1.** *Consider the following factorizations:*

$$P = \frac{n_p}{d_p} = \frac{n_p^+ n_p^-}{d_p^+ d_p^-} \qquad W = \frac{n_w}{d_w} = \frac{n_w}{d_w' d_p^+} \tag{5.8}$$

*where $n_p^+, d_p^+$ contain the unstable (or slow in the case of $d_p^+$) zeros of $n_p, d_p$, respectively. Similarly, $n_p^-, d_p^-$ contain the stable zeros of $n_p, d_p$. Then,*

$$S = \mathcal{N}^o W^{-1} = \rho \frac{q(-s)d_w}{q(s)n_w} \tag{5.9}$$

$$T = 1 - \mathcal{N}^o W^{-1} = \frac{n_p^+ \chi}{q(s)n_w} \tag{5.10}$$

$$K = \left( \frac{1 - \mathcal{N}^o W^{-1}}{\mathcal{N}^o W^{-1}} \right) P^{-1} = \frac{d_p^- \chi}{\rho n_p^- q(-s)d_w'} \tag{5.11}$$

*where $\chi$ is a polynomial such that*

$$q(s)n_w - \rho q(-s)d_w = n_p^+ \chi \tag{5.12}$$

*Proof.* The optimal weighted sensitivity $\mathcal{N}^o$ corresponds to

$$S = \mathcal{N}^o W^{-1} \quad \text{and} \quad T = 1 - \mathcal{N}^o W^{-1} \tag{5.13}$$

From the definitions of $S$ and $T$, the feedback controller can be expressed as

$$K = \frac{T}{S} P^{-1} = \frac{1 - \mathcal{N}^o W^{-1}}{\mathcal{N}^o W^{-1}} P^{-1} \tag{5.14}$$

Furthermore, the interpolation constraints (5.3) guarantee that $Q_0 \in \mathcal{RH}_\infty$. Thus, there exists a polynomial $\chi$ such that (5.6) can be rewritten as

$$Q_0 = \frac{d_p}{n_p^+ n_p^-} \left( \frac{q(s)n_w - \rho q(-s)d_w}{q(s)n_w} \right) = \frac{d_p \chi}{n_p^- q(s)n_w} \tag{5.15}$$

where the factorizations in (5.8) have been used. In terms of $Q_0$, we have that $S = 1 - PQ_0$, $T = PQ_0$ and $K = \frac{Q_0}{1 - PQ_0}$. Finally, straightforward algebra yields the polynomial structure of equations (5.9)–(5.11). $\square$

**Remark 5.2.1.** *It is noteworthy that the feedback controller (5.11) is realizable only if $P$ is biproper. Hence, in practice, it may be necessary to add fictitious high-frequency zeros to the initial model to meet this requirement.*

## 5.3   Weight selection

Let us denote by $\tau_1, \ldots, \tau_k$ the time constants of the unstable or slow poles of $P$. Equation (5.9) reveals that, except by the factor $\rho$, $|S|$ is determined by $|W^{-1}|$ ($\mathcal{N}^o$ is all-pass). Based on (5.9) and (5.10), the following structure for the weight is proposed

$$W = \frac{(\lambda s + 1)(\gamma_1 s + 1) \cdots (\gamma_k s + 1)}{s(\tau_1 s + 1) \cdots (\tau_k s + 1)} \tag{5.16}$$

where $\lambda > 0$, and

$$\gamma_i \in \big[\lambda, |\tau_i|\big] \tag{5.17}$$

The rationale behind the choice of $W$ in (5.16) is further explained below:

- Let us start assuming that $k = 0$ (i.e., $W = \frac{\lambda s + 1}{s}$). The integrator in $W$ forces $S(0) = 0$ for integral action. From (5.10), the term $(\lambda s + 1)$ in the numerator of $W$ appears in the denominator of the input-to-output transfer function. Consequently, the closed-loop will have a pole $s = -1/\lambda$. The idea is to use $\lambda$ to determine the speed of response, as in standard IMC.

- If $P$ has slow stable poles, it is necessary that $S$ cancels them if disturbance rejection is the main concern. Otherwise, they will appear in the transfer function $T_{yd} = SP$, making the response sluggish. This is why $W$ also contains these poles. As a result, slow (stable) and unstable poles are treated basically in the same way. This unified treatment ensures internal stability in terms of the generalized $\mathcal{D}$-stability region of Figure 5.1.

- As it has been said, producing $S(-1/\tau_i) = 0$, $i = 1, \ldots, k$ is necessary for internal stability and disturbance rejection. Notice, however, that these constraints mean decreasing $|S|$ at low frequencies. By a waterbed effect argument [72], recall the Bode's sensitivity integral:

$$\int_0^\infty |S(j\omega)|d\omega = \pi \sum_{i,\,\tau_i < 0}^{k} |\tau_i|^{-1} \tag{5.18}$$

This will augment $|S|$ at high frequencies, maybe yielding an undesirable peak ($M_S$) on it. This, in turn, will probably augment the peak of $|T|$ ($M_T$) and the overshoot in the

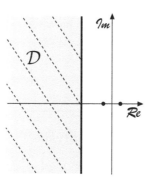

**FIGURE 5.1**
General stability region: slow and unstable poles are $\mathcal{D}$-unstable.

set-point response. In order to alleviate these negative effects, for each slow/unstable pole of $P$, we introduce a factor $(\gamma_i s + 1)$ in the numerator of $W$: as $\gamma_i \nearrow |\tau_i|$, $\left|\frac{\tau_i j\omega + 1}{\gamma_i j\omega + 1}\right| \searrow 1$; the resulting flatter frequency response will reduce the overshoot (improving the robustness properties, see Section 5.4) at the expense of settling time.

- We have supposed that $\lambda < |\tau_i| \ \forall k = 1 \ldots k$. In other words, we are considering relatively slow plants: for stable plants without slow poles, the standard IMC procedure will provide good results in terms of tracking and disturbance rejection; there is no conflict between $T_{yr}$ and $T_{yd}$. Note, in addition, that forcing $S = 0$ ($T = 1$) at high frequency is undesirable from a robustness point of view. This is why we discard rapid stable poles from the denominator of $W$. If the plant is unstable, there is no option and one has to force $S = 0$ ($T = 1$) at the rapid unstable poles, which imposes a minimum closed-loop bandwidth.

Essentially, there are two tuning parameters in $W$: $\lambda$ is intended to tune the robustness/performance compromise. The set of numbers $\gamma_i$ allows us to balance the performance between the servo and regulator modes. The latter point can be interpreted in terms of a mixed $S/SP$ sensitivity design: let us assume that $\lambda \approx 0$. Then, when $\gamma_i = |\tau_i|$ (servo tuning), we have that $|WS| \approx |S/s|$ and we are minimizing the peak of $|S|$ ($= |T_{er}|$) subject to integral action. In the other extreme, if $\gamma_i = \lambda$ (regulator tuning), the poles of $P$ appear in $W$. If the zeros of $P$ are sufficiently far from the origin, we have that $|WS| \approx |SP/s|$ in the low-middle frequencies. Heuristically, we are minimizing the peak of $|SP|$ ($= |T_{yd}|$) subject to integral action.

**Remark 5.3.1.** *Let us consider that $P$ has a RHP pole at $s = -1/\tau_i$ ($\tau_i < 0$) and a RHP zero at $s = z_i$. Then, from (5.3) and (5.16), it follows that*

$$\left|\frac{1}{\tau_i z_i + 1}\right| \left|\frac{(\lambda z_j + 1)\prod_{j=1}^{k}(\gamma_j z_j + 1)}{z_j \prod_{j=1, j\neq i}^{k}(\tau_j z_j + 1)}\right| = |\rho|\left|\frac{q(-z_i)}{q(z_i)}\right| \tag{5.19}$$

*As the RHP pole $-1/\tau_i$ and the RHP zero $z_i$ get closer to each other, $\tau_i z_i \to -1$, which makes the left-hand side grow unbounded. Since $\left|\frac{q(-z_i)}{q(z_i)}\right| \leqslant 1$, $|\rho| \to \infty$. Note that this happens regardless of the values of $\lambda$ and the $\gamma_j$'s, and it obeys the fact that plants with unstable poles and zeros close to each other are intrinsically difficult to control [51][72, Section 5.3].*

**Remark 5.3.2.** *For simplicity, the $\gamma_i$ parameters could be determined from a single parameter $\gamma \in [0, 1]$ as indicated below:*

$$(\gamma_1, \ldots, \gamma_k)^T = (1 - \gamma)(\lambda, \ldots, \lambda)^T + \gamma(|\tau_1|, \ldots, |\tau_k|)^T \tag{5.20}$$

## 5.4  Stability and robustness analysis

Because of the assumptions in Theorem 5.2.1, the possible delay of the plant must be approximated by a non-minimum phase rational term. This approximation creates a mismatch between $P$ (the purely rational model used for design) and the nominal model containing the time delay, let us call it $P_o$. The following sufficient condition for nominal stability can be derived from the conventional Nyquist stability criterion [72]:

**Proposition 5.4.1.** *Assume that $P$ is internally stabilized by $K$, and that $P$ and $P_o$ have the same RHP poles. Then, $K$ internally stabilizes $P_o$ if*

$$\left|\frac{L_o - L}{1 + L}\right| < 1 \ \forall\omega \in \Omega_{pc} \tag{5.21}$$

*where $L = PK, L_o = P_oK$, and $\Omega_{pc} = \left\{\omega : \angle\left(\frac{L_o - L}{1+L}\right) = -\pi + 2\pi n, \ n \in \mathbb{Z}\right\}$ is the set of phase crossover frequencies of $\frac{L_o - L}{1+L}$.*

Figure 5.2 illustrates the situation graphically for a stable plant: the distance from $L$ to the point $(-1, 0)$ must exceed $|L_o - L|$ when the *vectors* $L_o - L$ and $-1 - PK$ are aligned. Rather than using Proposition 5.4.1, a more practical approach is to check robust stability with respect to $P$ [72, 49], including $P_o$ in the uncertain set under consideration [81]. Generally, the way in which $\lambda$ and $\gamma_i$ influence robustness is:

- Augmenting $\lambda$ decreases the closed-loop bandwidth, making the system more robust and less sensitive to noise.

- Decreasing $\gamma_i$ improves the disturbance rejection, but increases the overshoot in the set-point response to the detriment of robustness.

These robustness implications can be understood in terms of the robust stability condition $\|\Delta T\|_\infty < 1$ (equivalently $|T| < 1/|\Delta| \ \forall\omega$), where $\Delta$ models the multiplicative plant uncertainty [72]. Augmenting $\lambda$ makes the system slower, which favours robust stability. On the other hand, decreasing $\gamma_i$ increases the peak of $|T|$ (responsible for the overshoot increment), which limits the amount of multiplicative uncertainty.

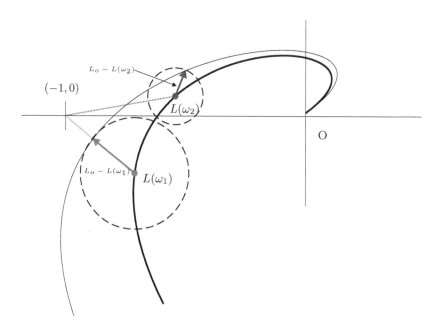

**FIGURE 5.2**
Stability condition for $P_o$ in terms of $P$. The magnitude condition $|1 + L| > |L_o - L|$ must be true for $\omega_1$ (which is a phase crossover frequency, i.e., $\omega_1 \in \Omega_{pc}$), but not for $\omega_2$.

## 5.5   Application to PI tuning

This section deals with the application of the presented design method to the tuning of PI compensators.

### 5.5.1   Stable/unstable plants

Let us consider the FOPTD model given by $P_o = K_g \frac{e^{-sh}}{\tau s+1}$, where $K_g, h, \tau$ are, respectively, the gain, the (apparent) delay, and the time constant—negative in the unstable case—of the process. For design purposes, we take

$$P = K_g \frac{-sh+1}{\tau s+1} \tag{5.22}$$

where a first-order Taylor expansion has been used to approximate the time delay. From (5.16) and (5.22), with $k = 1$, the following weight results

$$W = \frac{(\lambda s + 1)(\gamma s + 1)}{s(\tau s + 1)} \tag{5.23}$$

where $\lambda > 0, \gamma \in [\lambda, |\tau|]$. The optimal weighted sensitivity is determined from (5.3). In this case, $P$ has a single RHP zero ($\nu = 1$), and $\mathcal{N}^o$ becomes

$$\mathcal{N}^o = \rho = \frac{(\lambda + h)(\gamma + h)}{\tau + h} \tag{5.24}$$

From (5.11), the controller is finally given by

$$K = \frac{\chi}{K_g \rho s} \tag{5.25}$$

where

$$\chi = \frac{\tau(h + \lambda + \gamma) - \lambda\gamma}{\tau + h} s + 1 \tag{5.26}$$

The feedback controller (5.25) can be cast into the PI structure:

$$K = K_c \left(1 + \frac{1}{T_i s}\right) \tag{5.27}$$

according to the tuning rule in the first row of Table 5.1.

Essentially, the trade-off between disturbance rejection and set-point tracking is controlled by $T_i$. This can be verified by considering the proposed PI settings for the extreme values of $\gamma$. This has been done in Table 5.2 for the stable plant case ($\tau > 0$).

Certainly, $T_i$ is the parameter which varies more with $\gamma$: $K_c$ varies from $\frac{1}{K_g}\frac{\tau}{\lambda+h}$ to $\frac{1}{K_g}\frac{\tau}{\lambda+h}\left(\frac{h+2\lambda-\lambda^2/\tau}{h+\lambda}\right)$ as $\gamma$ is decreased from $\tau$ to $\lambda$. This way, as we improve disturbance

**TABLE 5.1**

Proposed PI tuning rules.

| Model | $K_c$ | $T_i$ | |
|---|---|---|---|
| $K_g \frac{e^{-sh}}{\tau s+1}$ | $\frac{1}{K_g}\frac{T_i}{\lambda+\gamma+h-T_i}$ | $\frac{\tau(h+\lambda+\gamma)-\lambda\gamma}{\tau+h}$ | $\lambda > 0, \gamma \in [\lambda, |\tau|]$ |
| $K_g \frac{e^{-sh}}{s}$ | $\frac{1}{K_g}\frac{T_i}{\lambda\gamma+hT_i}$ | $h + \lambda + \gamma$ | $\lambda > 0, \gamma \in [\lambda, \infty)$ |

**TABLE 5.2**
PI tuning rules for the extreme values of $\gamma$.

| | $\gamma = \lambda$ | | $\gamma = \tau$ | |
|---|---|---|---|---|
| $K_c$ | | $T_i$ | $K_c$ | $T_i$ |
| $\frac{1}{K_g}\frac{\tau}{\lambda+h}\left(\frac{h+2\lambda-\lambda^2/\tau}{h+\lambda}\right)$ | | $\frac{\tau(h+2\lambda)-\lambda^2}{\tau+h}$ | $\frac{1}{K_g}\frac{\tau}{\lambda+h}$ | $\tau$ |

rejection, the controller gain increases. The multiplicative factor $\frac{h+2\lambda-\lambda^2/\tau}{h+\lambda}$ equals one when $\lambda = \tau$. If $\tau \gg h, \lambda$, then $\frac{h+2\lambda-\lambda^2/\tau}{h+\lambda} \approx \frac{h+2\lambda}{h+\lambda} < 2$, which shows that $K_c$ augments moderately in the transition to the regulator mode. Based on these facts, it is reasonable to select $K_c = \frac{1}{K_g}\frac{\tau}{\lambda+h}$, and fix $T_i$ for good servo/regulation trade-off. This strategy is the essence of the SIMC tuning rule for stable plants [69].

Next, we will compare the input-to-output transfer functions achieved for the extreme values of $\gamma$. For small values of the time delay, $n_p^+ = -sh+1 \approx 1$, and equation (5.10) (with $q(s) = 1, \chi = \zeta s + 1, n_w = (\lambda s + 1)(\gamma s + 1)$) allows us to write:

$$|T(j\omega)| \approx \left|\frac{1}{\lambda j\omega + 1}\right|\left|\frac{\zeta j\omega + 1}{\gamma j\omega + 1}\right| \tag{5.28}$$

For a lag-dominant plant, the following approximations are valid:

- When $\gamma = \lambda$, the closed-loop magnitude is

$$|T(j\omega)| \approx \left|\frac{1}{\lambda j\omega + 1}\right|\left|\frac{\left(\frac{\tau(h+2\lambda)-\lambda^2}{\tau+h}\right)j\omega + 1}{\lambda j\omega + 1}\right| \approx \left|\frac{1}{\lambda j\omega + 1}\right|\left|\frac{(h + 2\lambda)j\omega + 1}{\lambda j\omega + 1}\right| \tag{5.29}$$

- When $\gamma = |\tau|$, we have that

$$|T(j\omega)| \approx \left|\frac{1}{\lambda j\omega + 1}\right| \tag{5.30}$$

for the stable plant case ($\tau > 0$). If $P$ is unstable ($\tau < 0$), $T$ is such that

$$|T(j\omega)| \approx \left|\frac{1}{\lambda j\omega + 1}\right|\left|\frac{\left(\frac{\tau(h+\lambda+|\tau|)-\lambda|\tau|}{\tau+h}\right)j\omega + 1}{|\tau|j\omega + 1}\right|$$

$$\approx \left|\frac{1}{\lambda j\omega + 1}\right|\left|\frac{(h + 2\lambda + |\tau|)j\omega + 1}{|\tau|j\omega + 1}\right| \tag{5.31}$$

Therefore, as the value of $\gamma$ is increased, the pole and the zero of $\frac{\zeta s+1}{\gamma s+1}$ in (5.28) get closer to each other, reducing the overshoot and providing flatter frequency response.

## 5.5.2 Integrating plant case ($\tau \to \infty$)

If the plant under control is integrating, it can be modelled by an integrator plus time delay (IPTD) model: $P_o = \frac{K_g e^{-sh}}{s}$. For this case, we take

$$P = K_g\frac{-sh + 1}{s} \tag{5.32}$$

The corresponding weight is chosen as

$$W = \frac{(\lambda s + 1)(\gamma s + 1)}{s^2} \tag{5.33}$$

where $\lambda > 0, \gamma \in [\lambda, \infty)$. The optimal weighted sensitivity becomes

$$\mathcal{N}^o = \rho = (\lambda + h)(\gamma + h) \tag{5.34}$$

From (5.11),

$$K = \frac{1}{K_g} \frac{\zeta' s + 1}{(\lambda\gamma + h\zeta')s} \tag{5.35}$$

where

$$\zeta' = h + \lambda + \gamma \tag{5.36}$$

The associated PI tuning rule can be consulted in the second row of Table 5.1. Alternatively, the tuning rules for the IPTD model could have been derived by taking the limit $\tau \to \infty$ in the FOPTD settings, considering the approximation $K_g \frac{e^{-sh}}{\tau s + 1} = \frac{K_g}{\tau} \frac{e^{-sh}}{s + 1/\tau} \approx \frac{K_g}{\tau} \frac{e^{-sh}}{s}$.

## 5.6  Simulation examples

This section evaluates the tuning rules given in Table 5.1 through four simulation examples. Examples 1–3 emphasize that the design presented in this chapter generalizes standard IMC. The purpose of the fourth example is to illustrate that, for simple plants and modest specifications, the presented design overcomes basic limitations of IMC, thus not being advisable to embark on more complex strategies. A summary of the controller settings for Examples 1–4 can be consulted in Table 5.3.

### 5.6.1  Example 1

The IMC-based PI tuning rule for stable FOPTD processes is given by [49]:

$$K_c = \frac{1}{K_g} \frac{\tau}{\lambda + h} \qquad T_i = \tau \tag{5.37}$$

In this example, the following concrete process $\frac{e^{-0.073s}}{1.073s + 1}$ is considered. Regarding the $\lambda$ parameter, two different values are chosen in order to achieve *smooth* ($\lambda = 0.10731$) and *tight* ($\lambda = 0.05402$) control [13], resulting in: $K_c^{sm} = 5.88, T_i^{sm} = 1.073$, and $K_c^{ti} = 8.38, T_i^{ti} = 1.073$. In the smooth control case, $M_S = 1.38$, whereas in the tight control case, $M_S = 1.71$. The associated disturbance responses are shown in Figure 5.3.

It is possible to reduce the magnitude of the disturbance rejection response by decreasing $\lambda$. However, the conventional IMC-based tuning continues to exhibit poor disturbance attenuation even for the tight case. To the detriment of robustness, decreasing further the value of $\lambda$ would improve the regulatory performance a little, but the response would continue to be sluggish. Accordingly, it is not possible to get both good regulatory performance and good robustness for the process under examination.

In the design of equation 5.5, setting $\gamma = \lambda$ produces an improvement of the regulation performance. Consequently, the problem reduces now to finding a value for $\lambda$ providing the prescribed robustness level. This is achieved for $\lambda = 0.1752$, which yields $M_S = 1.6551$. The corresponding time response is depicted in Figure 5.3.

It should be noted that the poor disturbance attenuation obtained through conventional IMC can be remedied in several (more *ad hoc*) ways. For example, approximating the process at hand by an integrating one [23]. Then, conventional IMC design gives

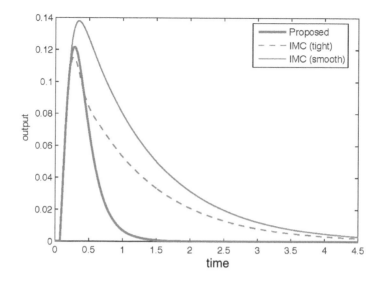

**FIGURE 5.3**
Load disturbance response for Example 1.

satisfactory disturbance rejection. A limitation of this approach is that it does not consider the servo/regulation trade-off. Other IMC-based approaches for improved regulatory performance can be found in [34, 64]. However, even for the simple FOPTD model, these approaches require a more complicated control structure (PID or PID plus filter). Overall, the presented tuning rules are simpler and more instructive.

### 5.6.2 Example 2

Generally speaking, the $\gamma$ parameter shifts the performance between setpoint tracking and disturbance rejection. To clarify this, we will continue Example 1, selecting $\lambda = 2h = 0.146$ and considering three different values for $\gamma$. The first value is $\gamma = \tau = 1.073$ (servo tuning). The resulting design is identical to the conventional IMC

**TABLE 5.3**
Tuning of $\lambda, \gamma$ (and corresponding PI settings) for Examples 1–4.

| Ex. | Model | $\lambda$ | $\gamma$ | $K_c$ | $T_i$ | Design type |
|---|---|---|---|---|---|---|
| 1 | $\frac{e^{-0.073s}}{1.073s+1}$ | 0.1752 | 0.1752 | 6.8765 | 0.3696 | Regulator |
| 2 | $\frac{e^{-0.073s}}{1.073s+1}$ | 0.146 | 1.073 | 4.8995 | 1.0730 | Servo (=IMC) |
| | | 0.146 | 0.4 | 5.8481 | 0.5286 | Servo/Regulation |
| | | 0.146 | 0.146 | 7.7215 | 0.3231 | Regulator |
| 3 | $\frac{e^{-s}}{-20s+1}$ | 2 | 2 | -11.56 | 5.4737 | Regulator ($\approx$IMC) |
| | | 0.9 | 9 | -11.9 | 11.9 | Servo/Regulation |
| 4 | $\frac{-1}{-s+1} \approx \frac{-e^{-0.01s}}{-s+1}$ | 0.1 | 0.1 | 18.2 | 0.22 | Regulator ($\approx$IMC) |
| | | 0.1 | 1 | 10.9 | 1.22 | Servo |
| | | 0.1 | 14 | 10.0642 | 15.667 | Servo ($K \approx 10$) |

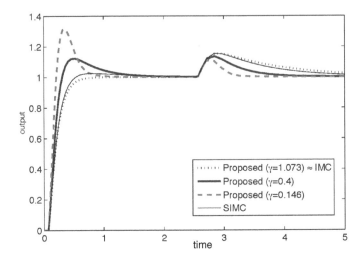

**FIGURE 5.4**
Tracking and disturbance responses for Example 2.

one. The second value is $\gamma = \lambda = 0.146$ (regulator tuning). Finally, we set $\gamma = 0.4$ for balanced servo/regulation performance.

Figure 5.4 shows the three time responses. We have also included the SIMC tuning rule [69]:

$$K_c = \frac{1}{K_g} \frac{\tau}{\lambda + h} \qquad T_i = \min\{\tau, 4(\lambda + h)\} \tag{5.38}$$

which was presented as a modification of the original settings (5.37) to improve the regulatory performance. Note, however, that in the *edge case* $\tau \approx 4(\lambda + h)$, there is no difference between (5.38) and (5.37). This is the situation in this example: $\tau = 1.073$ is close to $4(\lambda + h) = 0.876$. Looking at Figure 5.4, it is confirmed that the SIMC tuning gives approximately the same responses as conventional IMC. Lacking a rigorous analysis (this is not the intention here), the proposed PI tuning rule with $\gamma = 0.4$ seems to offer a better overall compromise. Finally, it is remarkable that, whereas the SIMC rule was derived only considering stable plants, the proposed tuning rule unifies the stable/unstable cases.

### 5.6.3   Example 3

For unstable plants, the IMC filter may cause large overshoot and poor robustness due to the large peak in the filter frequency response [22, 25]. The search of new filters to alleviate these shortcomings has resulted in more complicated (and application-specific) procedures [22]. In this example, we deal with an unstable plant, analyzing how the proposed method, albeit simple, can mitigate these negative effects. Let us consider the unstable process $\frac{e^{-s}}{-20s+1}$. The IMC controller is such that $T = e^{-s}f$, where $f = \frac{a_1 s+1}{(\lambda s+1)^2}$ and $a_1 = 20\left(e^{1/20}(\lambda/20+1)^2 - 1\right)$. Suppose that $\lambda = 2$ produces the desired closed-loop bandwidth, then $a_1 = 5.4408$. The feedback controller is $K = (-20s+1)\frac{f}{1-e^{-s}f}$, which is not purely rational. Approximating $e^{-s} \approx -sh + 1$, we finally obtain

$$K_{IMC} = \frac{-11.53s^2 - 1.542s + 0.1059}{s^2 - 0.04669s} \tag{5.39}$$

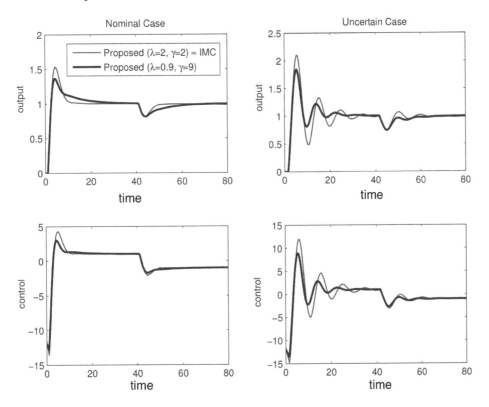

**FIGURE 5.5**
Tracking and disturbance responses for Example 3.

As for the proposed method, we start considering the initial tuning $\lambda = 2, \gamma = \lambda$. Figure 5.5 (nominal case) shows that this design is almost identical to the IMC one. Both $K_{IMC}$ and the proposed PI provide excellent disturbance rejection. However, it could be desirable to reduce the overshoot in the set-point response or improve the robustness properties. Within the IMC procedure, the only way to do it is to roll off the controller (increasing $\lambda$), making the system slower.

Contrary to this, if we take $\lambda = 0.9, \gamma = 9 \in [0.9, 20] = [\lambda, |\tau|]$, it can be seen from Figure 5.5 (nominal case) that it is possible to reduce the overshoot (at the expense of disturbance attenuation and settling time) without slowing down the system. Figure 5.6 depicts the frequency response of $|S|$ and $|T|$.

Recalling Section 5.4, the reduction of $M_S$ and $M_T$ confers more robustness and smoother control, as confirmed in Figure 5.5 (uncertain case), where the real plant delay is assumed to be $h = 1.6$ instead of one. Certainly, the new settings provide the best responses in both set-point tracking and disturbance attenuation.

### 5.6.4 Example 4

Finally, we revisit the design method in [25]. This $\mathcal{H}_\infty$ procedure was devised to generalize IMC: in particular, for unstable plants, it allows to use a different filter from that in (1.7), hence proving more flexible. The following design example, taken from [25], makes it clear: given the unstable plant $\frac{-1}{-s+1}$ ($P_a = 1, P_m = \frac{-1}{-s+1}$), the controller is designed in order to achieve a closed-loop response similar to $\frac{1}{0.1s+1}$ that corresponds to $f = \frac{1}{0.1s+1}$ in problem

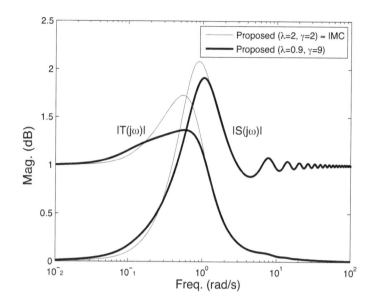

**FIGURE 5.6**
Magnitude frequency responses of $S$ and $T$ for Example 3. For $\lambda = 0.9, \gamma = 9$, the peaks of $|S|$ and $|T|$ are decreased without reducing the closed-loop bandwidth.

(1.10). This specification is coherent, in the sense that the desired closed-loop bandwidth is considerably beyond the unstable pole frequency [25]. Note that $P_a f|_{s=1,0} \approx 1$, taking into account internal stability constraints and zero steady-state error (unity low-frequency gain). The desired closeness between $T$ and $P_a f = \frac{1}{0.1s+1}$ is specified by the inequality $\|T - P_a f\|_\infty < \alpha$, with $\alpha = 0.1$. In addition, it is assumed that the actuators can pump up a maximum gain of 10 ($\beta_c = 10$). The frequency cost $\epsilon_1$ is chosen to gradually reach the maximum gain $\alpha/10$ as the plant model loses its bandwidth to the controller. Finally, $\epsilon_2 = 0$. Solving (1.10) leads to the $\mathcal{H}_\infty$ controller

$$K_\infty = \frac{1.099 \times 10^6 (s + 18.34)(s^2 + 6s + 9)}{(s + 1.15 \times 10^5)(s + 17.14)(s^2 + 5.94s + 8.85)} \qquad (5.40)$$

and the flag $\rho = 0.1 \leqslant \alpha$. This means that the desired objectives have been achieved. Figure 5.7 depicts the results both in the frequency and the time domain[1].

In view of Figure 5.7, it is clear that $K_\infty$ does not provide integral action, even if $f|_{s=0} = 1$. As claimed in [42], where this and other pitfalls in applying the design in [25] are highlighted, there are two possible sources of difficulty: first, the fact that $f|_{s=1}$ is not exactly one, as required by the unstable plant pole at $s = 1$. Second, the fact that $\epsilon_1 \neq 0$ or $\epsilon_2 \neq 0$, as it is also the case in this example.

In what follows, we will inspect the results obtained with the proposed method, leaving the $\lambda$ parameter fixed at $\lambda = 0.1$. Let us approximate $\frac{-1}{-s+1} \approx -\frac{e^{-0.01}}{-s+1}$ in order to apply the tuning rules of Table 5.1 . We start by selecting $\lambda = 0.1, \gamma = \lambda$, but the actuator limits are violated. In order to adhere to the given specifications, we take $\gamma = 1$, which almost verifies the actuator restriction. As a matter of fact, we can make the closed-loop closer to $f = \frac{1}{0.1s+1}$ by increasing further the value of $\gamma$ (the additional value $\gamma = 14$ has

---

[1]These plots are absent in [25].

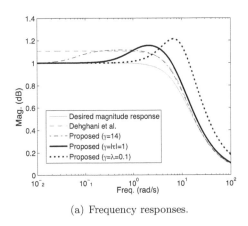

(a) Frequency responses.

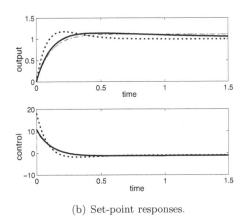

(b) Set-point responses.

**FIGURE 5.7**
Frequency and time responses for Example 4.

been considered). From Figure 5.7, it is evident that the proposed method always provides integral action. When $\gamma \to \infty$, a proportional controller $K = 10$ is obtained, for which the closed-loop is $\frac{1}{0.1s+0.9} \approx \frac{1}{0.1s+1}$. It is remarkable that $K_\infty$ can be handcrafted into such a plain gain too, yielding the same results as the original fourth-order controller. However, in [25], the application of a model-reduction algorithm only lowered the order of $K_\infty$ to three. This point stresses that care has to be taken when using/implementing numerical designs. For the particular case at hand, $\gamma = 1$ gives a compromise between the desired magnitude response, control effort, controller complexity, and the inclusion of integral action in the loop. Obviously, the proposed design may be insufficient for more stringent specifications. In these cases, the more flexible procedure in [25] reveals advantageous.

# 6

# Generalized IMC Design and $\mathcal{H}_2$ Approach

To cope with the input/output *regulator* problem, we rely here on the IMC paradigm [49] already presented in previous chapters. Based on IMC, we present a design method to take into account both input and output disturbances. The proposed design provides generalized IMC filters that can be used to obtain good results in terms of output sensitivity (favouring output disturbances), or in terms of input sensitivity (therefore placing the emphasis on load disturbances). If both input and output disturbances are expected, the design offers the possibility of obtaining a balance that improves the overall disturbance rejection response.

The analytical solution presented here can be seen as the $\mathcal{H}_2$ counterpart of what has been presented so far. With respect to the approach developed in the previous chapter, some assumptions have been removed: i.e., the plant model is not restricted to be purely rational nor to contain at least one right half-plane (RHP) zero. In addition, plants with complex poles have been included in the discussion. An interesting aspect of the here-adopted $\mathcal{H}_2$ approach is that it unifies previous designs that can also be found in the literature as in [22, 34, 43], resulting into a more general structure for the IMC filter. The distinguishing feature of the new filter is that it allows to balance the input/output regulatory performance in a simple manner. This is a fundamental trade-off, disregarded in [22, 34, 43], that cannot be overcome using a 2DOF control configuration or a related approach as done in the works [87, 64, 65].

## 6.1 Motivation for the input/output disturbance trade-off

The objective of a control system is to make the output $y$ behave in a desired way by manipulating the plant input $u$. There are basically two different problems [58, 72]: the *servo* problem, which concerns the tracking of the reference signal $r$, and the *regulator* problem, which aims at rejecting the disturbances $d$ entering the control loop. In both cases, the controller $K$ is designed to make the control error $e = y - r$ small. Here we do concentrate exclusively with the regulator problem. Note that if the resulting tracking performance was not suitable, this could be fixed in a second step by introducing a reference prefilter [49, 72]. More generally, the servo and the regulator problems can be solved independently by using a 2DOF topology [58, 84, 35]. In what follows, we will assume that disturbances cannot be measured and that can enter both at the input and at the output of the plant $P$. Therefore, a feedforward strategy [27, 83] is not advantageous in the considered scenario, where the feedback controller completely determines the disturbance response.

Integrating (and close to integrating) processes are very common in industry (e.g., level systems and pulp and paper plants). For illustration purposes, let us consider a pure integrator process controlled by a proportional-integral (PI) compensator. The situation is depicted in Figure 6.1(a).

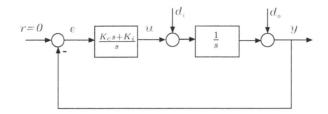

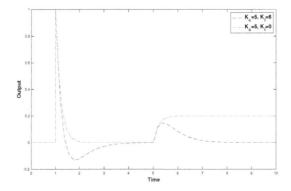

(a) Integrator process with PI controller.

(b) Time responses to unity step disturbances at the output ($t = 1$)
and at the input ($t = 5$) of the plant. $K = \frac{K_c s + K_i s}{s}$

**FIGURE 6.1**
Motivating example.

As it can be seen in Figure 6.1(b), a proportional controller ($K_c = 5, K_i = 0$) yields excellent results when output disturbances are the main concern. However, the corresponding response to load disturbances is not satisfactory. In order to suppress the steady-state error due to input disturbances, integral action is necessary. The response to the alternative settings $K_c = 5, K_i = 6$ confirms this point. It is clear from Figure 6.1(b) that there exists a fundamental trade-off between input and output disturbances. Therefore, if both kind of disturbances are present, a trade-off tuning methodology would help improve the overall disturbance response. These issues concerning input/output disturbances are the main concern of this chapter.

Historically, the inherent shortcomings of the IMC method have resulted in the search of new filters and/or alternative procedures: for minimum-phase (MP) unstable plants, Campi [22] suggested a filter which allows easy adjustment of the closed-loop bandwidth as well as a robustness improvement. For stable plants, in [34] a modified version of the the conventional filter for enhanced input disturbance attenuation is presented. From a broader viewpoint, a simple IMC-based procedure applicable to both stable and unstable plants and aimed at input disturbances was presented in [43]. Some years later, Dehghani, in [25], reported the difficulties with the IMC procedure in an exhaustive manner and, in order to undergo them, devised a numerical design blending IMC and $\mathcal{H}_\infty$ ideas. Although the latter design offers great versatility, it requires judicious choices for some frequency weights and for the desired closed-loop response, which may lead to design pitfalls as noted in [42].

## 6.2 Problem statement

To set the problem, we make use of Single-Input Single-Output (SISO) linear models of the form

$$y = Pu + Wd \tag{6.1}$$

for which the corresponding feedback setup is depicted in Figure 6.2. In (6.1), $W$ (which is not a physical component as $P$ or $K$) represents a frequency weight that will be designed to make it easy to balance the disturbance response at the input and at the output of the plant. By absorbing the input type (e.g. step-like disturbances) into the weight $W$ too, we will assume hereafter that $d$ in figure 6.2 is an impulse, i.e. $d = 1$. To derive the feedback controller $K$, we will look at the structure of $\mathcal{H}_2$-optimal controllers. By an $\mathcal{H}_2$-optimal controller, we understand here one such that the integrated square error

$$\|e\|_2^2 = \int_0^\infty e^2(t)dt \tag{6.2}$$

is minimized for a particular input. Bearing in mind that $e = -SWd = -SW$, where $S \doteq \frac{1}{1+PK}$ is the sensitivity function, we can state problem (6.2) in the frequency domain

$$\min_{\mathcal{C}}\|e\|_2^2 = \min_{\mathcal{C}}\frac{1}{2\pi}\int_{-\infty}^{\infty}|S(j\omega)W(j\omega)|^2 d\omega \tag{6.3}$$

where $\mathcal{C}$ denotes the set of *internally stabilizing* controllers. Internal stability is the requirement that all the closed-loop transfer functions are stable, implying that cancellation of unstable poles between the plant $P$ and the controller $K$ is not allowed. It is well-known that the IMC parameterization of the feedback controller[49],

$$K = \frac{Q}{1 - PQ}, \tag{6.4}$$

allows writing all the closed-loop relations affinely in $Q$ (e.g., $S = 1 - PQ, T = PQ$). Then, in terms of $Q$, the following fundamental result solves (6.3):

**Theorem 6.2.1** (Morari and Zafiriou [49]). *Let us factor both the plant $P$ and the weight $W$ into an all-pass and a minimum phase (MP) portion so that $P = P_a P_m$ and $W = W_a W_m$.*

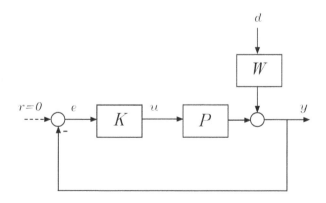

**FIGURE 6.2**
Basic setup for the input/output regulator problem.

*Denote by $l, k$ the number of integrators and unstable poles of $P$, respectively. Now, assume that the weight $W$ contains $l' \geq l$ integrators and the first $0 \leq k' \leq k$ unstable poles of $P$ and define:*

$$b_P = \prod_{i=1}^{k} \frac{-s + \pi_i}{s + \bar{\pi}_i} \quad \text{and} \quad b_W = \prod_{i=1}^{k'} \frac{-s + \pi_i}{s + \bar{\pi}_i}, \tag{6.5}$$

*being $\pi_1, \cdots, \pi_k$ are the unstable poles of $P$. Then, the $\mathcal{H}_2$-optimal (internally stabilizing) $Q$ is given by*

$$Q = b_P (P_m b_W W_m)^{-1} \left\{ (b_P P_a)^{-1} b_W W_m \right\}_*  \tag{6.6}$$

*where the operator $\{\}_*$ denotes that after a partial fraction expansion (PFE) of the operand all terms involving the poles of $P_a^{-1}$ are omitted.*

Note that it is straightforward how to select $W$ for the extreme cases at hand. For example, if only step output disturbances are considered, the weight should be $W = 1/s$, whereas $W = P/s$ for the case of step disturbances entering at the input of the plant. A more difficult problem is how to select $W$ systematically for balanced operation. In addition, $W$ should be such that it allows adjusting the robustness/performance trade-off. The selection of $W$ will be fully addressed in Section 3. We end this section by observing the following facts:

**Remark 6.2.1.** *The optimal solution in (6.6) only depends on the MP part of $W$. Consequently, $W$ can be restricted to be MP without loss of generality (i.e., $W = W_m$).*

**Remark 6.2.2.** *For MP (possibly unstable) plants ($P_a = 1$), the optimal solution in (6.6) becomes $Q = P_m^{-1}$, independently of $W$.*

## 6.3   Weight selection

This section first addresses the selection of a suitable weight $W$ for the problem at hand. After selecting $W$, an analytical solution for $Q$ is given based on the $\mathcal{H}_2$ minimization criterion and, finally, we will examine the nominal performance, robust stability, and robust performance properties of the derived controller.

Let us take $P = P_a P_m$ as in Section 6.2, and denote by $d_d$ the generating polynomial of the disturbance (i.e., $d_d = s$ for steps, $d_d = s^2$ for ramps, etc.). For the sake of clarity, it is temporarily assumed that $P_a \neq 1$ and that $P$ has no complex poles or zeros, nor any pole at the origin. We also assume that $P$ has slow/unstable poles at $s = -1/\tau_1, \ldots, -1/\tau_k$. Then, we make the following choice of the weight:

$$W = \frac{(\lambda s + 1)^n}{d_d} \prod_{i=1}^{k} \frac{\gamma_i s + 1}{\tau_i s + 1} \tag{6.7}$$

with

$$n = \max \left\{ 1, \delta(d_d) + \delta(P) - 1 \right\} \tag{6.8}$$

where $\delta(d_d), \delta(P)$ denote the degree of $d_d$ and the relative degree of $P$, respectively. For the common case of step disturbances ($d_d = s$), (6.8) simplifies to $n = \max \left\{ 1, \delta(P) \right\}$. Finally, $\lambda$ and $\gamma_1, \ldots, \gamma_k$ in (6.7) are tuning parameters verifying that

$$\lambda > 0, \qquad \gamma_i \in [\lambda, |\tau_i|] \tag{6.9}$$

We recall here that the main objective of our design is to consider disturbances entering both at the input and at the output of the plant. In addition, the design has to account for model uncertainty. The rationale behind the selection of $W$ in (6.7) is explained below:

- In order to explain the role of $\lambda$ and $\gamma_i$ separately, let us start considering that $\lambda = 0$. Then, we have that $W = \frac{1}{d_d} \prod_{i=1}^{k} \frac{\gamma_i s+1}{\tau_i s+1}$. Now, by making $\gamma_i = \lambda = 0, i = 1, \ldots, k$, the weight is $W = \frac{1}{d_d} \prod_{i=1}^{k} \frac{1}{\tau_i s+1}$. For this choice of $\gamma_i$, the design will provide good results in terms of input disturbance attenuation since we are including the slow/unstables poles of $P$ in $W$. Stated otherwise, the disturbance passes through the conflicting poles of the plant (note that fast, stable poles do not impose a trade-off between input/output regulatory performance).

- At this point, we can improve the output disturbance response by increasing the value of each $\gamma_i$. To see this, let us consider that $\gamma_i$ is set to the upper bound of the interval (6.9), i.e., we take $\gamma_i = |\tau_i|$. It is then clear that $|W| = \frac{1}{|d_d|}$, for which (6.3) optimizes the ISE for output disturbances.

- So far, we have assumed that $\lambda = 0$. Let us suppose now that each $\gamma_i$ has been fixed to a particular value. As we increase the value of $\lambda$, the minimization in (6.3) will penalize the magnitude of $S$ at higher frequencies, resulting into a slower closed-loop. Therefore, $\lambda$ can be used to adjust the robustness/performance trade-off. Regarding $n$, the value in (6.8) will ensure the properness of the final controller (this point will be clarified later).

**Remark 6.3.1.** *For simplicity, the $\gamma_i$ parameters could be determined from a single parameter $\gamma \in [0,1]$ as indicated below:*

$$(\gamma_1, \ldots, \gamma_k)^T = (1-\gamma)(\lambda, \ldots, \lambda)^T + \gamma(|\tau_1|, \ldots, |\tau_k|)^T \tag{6.10}$$

## 6.4 Analytical solution

The next step towards obtaining the IMC controller is to use Theorem 6.2.1. As $W = W_m$ in (6.7) contains the unstable poles of $P$, we have that $b_P = b_W$, and (6.6) simplifies to $Q = (P_m W)^{-1} \{P_a^{-1} W\}_*$. This is a valid controller when $P$ is nonminimum-phase (NMP) in the sense that it is internally stabilizing and proper. However, recalling Remark 6.2.2, for MP plants (i.e., $P_a = 1$) the solution is $Q = P_m^{-1}$ regardless $W$. As a consequence, $Q$ may be improper, and it would be necessary to extend $Q$ by cascading a filter as in the conventional IMC procedure. We want to avoid this approach and obtain a proper solution directly from the specified weight $W$. Towards this objective, we finally propose the following solution

$$Q = (P_m W)^{-1} \left\{ P_a^{-1} W \right\}_\star \tag{6.11}$$

where $\{\cdot\}_\star$ acts like $\{\cdot\}_*$, when the plant is NMP; when $P$ is MP, the actuation of $\{\cdot\}_\star$ can be understood in terms of $\{\cdot\}_*$ as follows:

$$\left\{ P_a^{-1} W \right\}_\star \doteq \left\{ \left( \underbrace{P_a e^{-sh}}_{P_a'} \right)^{-1} W \right\}_* \Bigg|_{h=0} \tag{6.12}$$

That is to say, we consider a fictitious delay $h$, apply $\{\cdot\}_*$ and then evaluate at $h = 0$. The following example illustrates how to calculate (6.11):

**Example 6.4.1.** *Let us consider the (possibly unstable) FOPTD model $P = K_g \frac{e^{-sh}}{\tau s + 1}$, for which $P_m = \frac{K_g}{\tau s + 1}, P_a = e^{-sh}$. We assume that $|\tau| \gg h > 0$ ($k = 1$). In addition, we assume step-like disturbances, i.e., $d_d = s$, and take $n = 1$. By substituting into (6.7), we get $W = \frac{(\lambda s + 1)(\gamma s + 1)}{s(\tau s + 1)}$, with $\lambda > 0, \gamma \in [\lambda, |\tau|]$. If $h > 0$, the proposed controller (6.11) is identical to $\mathcal{H}_2$-optimal one:*

$$
\begin{aligned}
Q &= \frac{(\tau s + 1)^2 s}{K_g(\lambda s + 1)(\gamma s + 1)} \left\{ e^{sh} \frac{(\lambda s + 1)(\gamma s + 1)}{s(\tau s + 1)} \right\}_* \\
&= \frac{(\tau s + 1)^2 s}{K_g(\lambda s + 1)(\gamma s + 1)} \left\{ e^{sh} \frac{(\lambda s + 1)(\gamma s + 1)}{s(\tau s + 1)} \right\}_* \\
&= \frac{(\tau s + 1)^2 s}{K_g(\lambda s + 1)(\gamma s + 1)} \left( \frac{1}{s} - \frac{\tau e^{-h/\tau} \tau (1 - \lambda/\tau)(1 - \gamma/\tau)}{\tau s + 1} \right) \\
&= \frac{(\tau s + 1)([\tau - \tau e^{-h/\tau}(1 - \lambda/\tau)(1 - \gamma/\tau)]s + 1)}{K_g(\lambda s + 1)(\gamma s + 1)} \quad (6.13)
\end{aligned}
$$

*In the case $h = 0$, $P$ becomes MP ($P_a = 1$). In this case, we make $h = 0$ in (6.13), and we arrive at*

$$
Q = \frac{(\tau s + 1)([\tau - \tau(1 - \lambda/\tau)(1 - \gamma/\tau)]s + 1)}{K_g(\lambda s + 1)(\gamma s + 1)} \quad (6.14)
$$

*In particular, note that $Q \to P_m^{-1} = (\tau s + 1)/K_g$ as $\lambda \to 0$, implying that the $\mathcal{H}_2$-optimal solution is approached for small values of $\lambda$.*

The following proposition summarizes the most basic properties of the proposed controller[1]:

**Proposition 6.4.1.** *The IMC controller $Q$ in (6.11) is such that:*

*(P1) $Q$ is $\mathcal{H}_2$-optimal if $P$ is NMP. In the MP case, $Q$ tends to be $\mathcal{H}_2$-optimal when $\lambda \to 0$ provided that $\delta(d_d) \geq 1$.*

*(P1) $Q$ is proper and stable.*

*(P3) $S = 1 - PQ = 0$ at the poles of $W$.*

*Proof.*

(P1) The $\mathcal{H}_2$-optimal solution is given by $Q = (P_m W)^{-1} \{P_a^{-1} W\}_*$. The difference with respect to the proposed solution amounts to the $\{\cdot\}_*$ operator. If $P_a \neq 1$, $\{\cdot\}_*$ and $\{\cdot\}_*$ coincide because $\{P_a^{-1} W\}_*$ is strictly proper. Thus, in this case, (6.11) is optimal (with respect to the selected $W$). When $P_a = 1$, $\{P_a^{-1} W\}_* \to \{P_a^{-1} W\}_*$ when $\lambda \to 0$. This is because $P_a^{-1} W = W$ tends to become strictly proper (the $n$ zeros at $s = -1/\lambda$ of $W$ move to infinity). By definition of $\{\cdot\}_*, \{\cdot\}_*$, both yield the same result when applied to strictly-proper operands. Thus, the proposed $Q$ tends to $Q = (P_m W)^{-1} \{P_a^{-1} W\}_* = P_m^{-1}$ when $\lambda \to 0$.

(P2) From (6.7) and (6.11), straightforward algebra shows that the structure of $Q$ is given by

$$
Q = \frac{P_m^{-1} \chi}{(\lambda s + 1)^n (\gamma_1 s + 1) \cdots (\gamma_k s + 1)} \quad (6.15)
$$

---

[1] Although we are assuming that $W$ is given by (6.7), property (P3) holds generally for any MP weight.

where $\chi$ is a polynomial of degree $\delta_d(d_d) + k - 1$. Therefore, $\delta(Q) = n - \delta(P) - \delta(d_d) + 1$. Selecting $n$ as in (6.8) provides $\delta(Q) \geq 0$, implying that $Q$ is proper. Stability is also easy to check: the poles of $Q$ are the left half-plane (LHP) zeros of $P$, collected in $P_m$, and the zeros of $W$, which are also in the LHP.

(P3) Equivalently, we will show that $T = 1 - S = 1$ at the poles of $W$. The complementary sensitivity function is

$$T = PQ = (P_a^{-1}W)^{-1}\left\{P_a^{-1}W\right\}_\star \tag{6.16}$$

If $W$ has a pole at $s = p$ of multiplicity $m$, then we can write $P_a^{-1}W = \frac{\phi(s)}{(s-p)^m}$, and (6.16) can be expressed as

$$T = \frac{(s-p)^m}{\phi(s)}\left(\cdots + \sum_{i=1}^{m-1}\frac{\alpha_i}{(s-p)^i} + \frac{\alpha_m}{(s-p)^m} + \cdots\right) \tag{6.17}$$

where $\alpha_m = \phi(p)$. It is clear then that $T|_{s=p} = \left(\frac{\alpha_m}{\phi(s)}\right)\Big|_{s=p} = 1$.

$\square$

Property (P1) can be interpreted as the combination of the two steps of the IMC procedure into a single one. Properties (P2) and (P3) imply that $Q$ is realizable and internally stabilizing (because $W$ contains the poles of $P$). In addition, (P3) means asymptotic rejection of the disturbances (because the denominator of $W$ contains the generating polynomial $d_d$, recall the *internal model principle* [49, 72]).

**Remark 6.4.1.** *From (6.11) and property (P3), $Q$ and $1 - PQ$ have zeros at the $k$ slow/unstable poles of $P$. These zeros get cancelled when forming the equivalent unity feedback controller $K = \frac{Q}{1-PQ}$. This means that adjusting $\lambda, \gamma_i$ in $W$ does not change the structure of the final controller but only its parameters.*

**Remark 6.4.2.** *Strictly speaking, properties (P2) and (P3) are not sufficient conditions for internal stability when $P$ is a delayed unstable system. As explained in [91], in this case there are irremovable RHP pole/zero cancellations in $K$ that do not allow a direct implementation. In general, $Q$ can be approximated by a practical controller $K$ (e.g., PID type) by following different methodologies [88, 64, 65].*

## 6.4.1 Interpretation in terms of alternative IMC filters

The proposed controller (6.11) can be expressed as

$$Q = P_m^{-1}f \tag{6.18}$$

with $f = W^{-1}\left\{P_a^{-1}W\right\}_\star$. Let us take $W = \frac{n_w}{d_w}$. Now, considering how $\{\cdot\}_\star$ acts and taking into account property (P3) in Proposition 3.1, we can alternatively express $f$ as

$$f = \frac{\chi}{n_w} = \frac{\sum_{i=0}^{\delta(d_w)-1} a_i s^i}{(\lambda s + 1)^n \prod_{i=1}^{k}(\gamma_i s + 1)} \tag{6.19}$$

where $a_0, \ldots, a_{\delta(d_w)-1}$ are determined from the following system of linear equations

$$T|_{s=\pi_i} = P_a f|_{s=\pi_i} = 1 \qquad i = 1 \ldots \delta(d_w) \tag{6.20}$$

being $\pi_i = -1/\tau_i, i = 1..\delta(d_w)$ the poles of $W$. From (6.7), $\delta(d_w) = k + \delta(d_d)$ in general, except when $P$ is stable and we take $\gamma_i = \tau_i$ for all $i$. In the latter case, the weight (6.7) simplifies to $W = \frac{(\lambda s+1)^n}{d_d}$, and $\delta(d_w) = \delta(d_d)$. Note that, as long as the $a_i$ coefficients satisfy (6.20), the filter time constants $\lambda$ and $\gamma_i$ can be selected freely without any concern for nominal stability. In more detail, (6.20) corresponds to

$$
\begin{pmatrix}
\pi_1^{\delta(d_w)-1} & \cdots & \pi_1 & 1 \\
\vdots & \ddots & \vdots & \vdots \\
\pi_{\delta(d_w)}^{\delta(d_w)-1} & \cdots & \pi_{\delta(d_w)} & 1
\end{pmatrix}
\begin{pmatrix}
a_{\delta(d_w)-1} \\
\vdots \\
a_0
\end{pmatrix}
=
\begin{pmatrix}
P_a^{-1} n_w|_{s=\pi_1} \\
\vdots \\
P_a^{-1} n_w|_{s=\pi_{\delta(d_w)}}
\end{pmatrix}
\tag{6.21}
$$

In the context of step-like inputs, the filter (6.19) generalizes some previously reported filters in the following way:

- For stable plants, by taking $\gamma_i = \tau_i$, the conventional IMC filter[49] is obtained. However, if we take $\gamma_i = \lambda$, then the filter in [34] results.

- Essentially, the filter suggested in [22] for MP unstable plants corresponds to taking $\gamma_i \to \infty$ in (6.19). In the general unstable plant case, the filter in [43] is recovered by choosing $\gamma_i = \lambda$.

Finally, by using Lagrange-type interpolation theory [49], it is possible to develop an expression[2] for (6.19) explicitly:

$$
f = \frac{1}{n_w} \sum_{j=1}^{\delta(d_w)} (P_a^{-1} n_w)|_{s=\pi_j} \prod_{\substack{i=1 \\ i \neq j}}^{\delta(d_w)} \frac{s - \pi_i}{\pi_j - \pi_i}
\tag{6.22}
$$

### 6.4.2  Extension to plants with integrators or complex poles

It has been shown in Section 6.4.1 that the proposed design amounts to the selection of a convenient filter $f$, so that $Q = P_m^{-1} f$. Here, we detail the structure of such a filter (passing over $W$ for brevity) when $P$ has integrators and/or complex conjugate poles. To keep it simple, we address each situation at a time:

(i) $P$ has (exclusively) $l$ poles at the origin

Then, the corresponding filter is

$$
f = \frac{\sum_{i=0}^{\delta(d_W)-1} a_i s^i}{(\lambda s + 1)^n \prod_{i=1}^{l}(\gamma_i s + 1)}
\tag{6.23}
$$

where $\delta(d_W) = l + \delta(d_d)$. The only difference with respect to (6.19) is that now $\gamma_i \in [\lambda, \infty)$, whereas for slow/unstable poles we had $\gamma_i \in [\lambda, |\tau_i|]$. This can be easily understood, since an integrator corresponds to a pole with an infinitely large time constant.

(ii) $P$ has (exclusively) $m$ complex conjugate poles

Let us suppose that the $m$ complex conjugate poles are at $-\xi_i \omega_{ni} \pm j\omega_{ni}\sqrt{1-\xi_i^2}$. Then, the structure of the filter is

$$
f = \frac{\sum_{i=0}^{\delta(d_W)-1} a_i s^i}{(\lambda s + 1)^n \prod_{i=1}^{m}(\gamma_{i,2} s^2 + \gamma_{i,1} s + 1)}
\tag{6.24}
$$

---

[2]The formula (6.22) is not valid for repeated poles.

where $\delta(d_W) = 2m + \delta(d_d)$. For input disturbances, $\gamma_{i,2} = \lambda^2, \gamma_{i,1} = 2\lambda$ so that $\prod_{i=1}^{m}(\gamma_{i,2}s^2+\gamma_{i,1}s+1) = (\lambda s+1)^{2m}$. For output disturbances, we want $\prod_{i=1}^{m}(\gamma_{i,2}s^2+\gamma_{i,1}s+1)$ equal to $(1/\omega_{ni}^2)\prod_{i=1}^{m}(s^2+2|\xi_i|\omega_{ni}+\omega_{ni}^2)$, which is achieved for $\gamma_{i,2} = (1/\omega_{ni})^2, \gamma_{i,1} = 2|\xi_i|/\omega_{ni}$ (in this extreme case, only when $P$ is stable, $\delta(d_W) = \delta(d_d)$ as explained in Section 3.2.1). It is not so simple now to determine an interval for $\gamma_{i,1}, \gamma_{i,2}$ as in the real poles' case (this point is illustrated in Section 6.4). An exception occurs if the complex poles are well-damped ($|\xi_i|$ close to one); in this case we can disregard the imaginary part and treat the complex conjugate pairs as double real poles at $s = -\omega_{ni}\xi_i$. This allows simplifying the filter structure to

$$f = \frac{\sum_{i=0}^{\delta(d_W)-1} a_i s^i}{(\lambda s + 1)^n \prod_{i=1}^{m}(\gamma_i s + 1)^2} \tag{6.25}$$

with $\gamma_i \in \left[\lambda, \frac{1}{|\omega_{ni}\xi_i|}\right]$.

## 6.5 Performance and robustness analysis

In any practical design method, robust performance is the ultimate goal: we want the controller to work well under uncertain circumstances. Assuming that a condition for robust stability is met, the next subsection gives an upper bound for the performance degradation with respect to the nominal case.

From Section 6.2, the ISE for an output disturbance $d = 1/d_d$ is given by

$$\text{ISE}_o = \int_0^{\infty} e^2(t)dt = \frac{1}{2\pi}\int_{-\infty}^{\infty}\left|Sd_d^{-1}(j\omega)\right|^2 d\omega \tag{6.26}$$

Similarly, when $d$ enters at the input of the plant, the corresponding ISE is

$$\text{ISE}_i = \int_0^{\infty} e^2(t)dt = \frac{1}{2\pi}\int_{-\infty}^{\infty}\left|PSd_d^{-1}(j\omega)\right|^2 d\omega \tag{6.27}$$

Equations (6.26) and (6.27) indicate the nominal performance achieved by the final design in terms of input/ouput disturbance attenuation. Robust stability can be assessed by the well-known condition [49, 72]

$$\|\Delta T\|_\infty = \sup_\omega |\Delta(\omega)T(j\omega)| < 1 \tag{6.28}$$

where $\Delta(\omega) \geq 0$ is a bound for the plant multiplicative uncertainty. In practice, nominal performance and robust stability alone are not enough because some plants in the uncertain set may be on the verge of instability, yielding very poor performance. It is therefore necessary to guarantee some degree of robust performance. To this aim, it is useful to have an upper bound for both $\text{ISE}_i$ and $\text{ISE}_o$. The worst error is generated by the worst plant, which can be expressed as $P(1 + \delta(s)\Delta(\omega))$ for some $\delta(s)$ such that $|\delta(j\omega)| \leq 1$. By using the inequality $|1 + P(1+\delta\Delta)K| \geq |1 + PK| - |PK|\Delta$, the actual sensitivity function $S$ can be bounded as

$$|S| = \left|\frac{1}{1 + P(1+\delta\Delta)K}\right| \leq \left|\frac{1}{1 - |\Delta T|}\right||S| \tag{6.29}$$

From (6.26), (6.27) and (6.29), the following upper bounds for the actual errors result in

$$\text{ISE}_o \leq \overline{\text{ISE}_o} = \frac{1}{2\pi}\int_{-\infty}^{\infty}\left|\frac{1}{1 - |\Delta T|}\right|^2 |Sd_d^{-1}|^2 d\omega \tag{6.30}$$

$$\text{ISE}_i \leq \overline{\text{ISE}_i} = \frac{1}{2\pi} \int_{-\infty}^{\infty} \left| \frac{1}{1 - |\Delta T|} \right|^2 |PSd_d^{-1}|^2 d\omega \tag{6.31}$$

As it is logical, the modeling error increases the (finite) gap between $\text{ISE}_i$ ($\text{ISE}_o$) and $\overline{\text{ISE}_i}$ ($\overline{\text{ISE}_o}$) as the stability boundary in (6.28) is approached, exhibiting the typical trade-off between nominal performance and performance degradation [49, 95, 72].

## 6.6   Tuning guidelines

In view of Equations (6.26), (6.27), (6.30), and (6.31), nominal performance is captured in terms of $S$, whereas robust performance is expressed using $T$ (which also determines robust stability (6.28)). The shape of these transfer functions depends on the values of the tuning parameters. The role of $\lambda$ is the same as in the conventional IMC: basically, for a given value of each $\gamma_i$, increasing $\lambda$ makes the system slower, to the detriment of $ISE_o$ and $ISE_i$, but favouring the robust stability condition (6.28) by reducing the closed-loop bandwidth. Let us consider now that $\lambda, \gamma_j, j = 1..k, j \neq i$ have been fixed, and see which is the influence of $\gamma_i$. From earlier discussion, when $\gamma_i = \lambda < |\tau_i|$, $W$ is asking for good load disturbance rejection by forcing $S = 0$ at $s = -1/\tau_i$, which may be responsible for a large peak on $|S|$ and $|T|$ and a somewhat aggressive response [10]. As we increase $\gamma_i$, $W$ specifies lower gains for $|S|$ at middle-high frequencies, which by a *waterbed effect* argument [72, 10] is achieved augmenting $|S|$ at low frequencies. Consequently, augmenting $\gamma_i$ has a *smoothing* effect. In particular, this means that improving the response to output disturbances will also make the system slower. As it will be shown in next section, after increasing $\gamma_i$, $\lambda$ can be decreased to compensate for the reduction of the closed-loop bandwidth. In summary, tuning $\gamma_i$ also has an effect on robustness, but it should be clear that the way of affecting the robustness properties is different: $\lambda$ is more related to the closed-loop bandwidth, which by the robust stability condition (6.28) is responsible for robustness in the high-frequency region (model uncertainty). On the other hand, the $\gamma_i$ parameters affect the mid-frequency robustness properties altering the peaks of the sensitivity functions. More precisely, augmenting $\gamma_i$ contributes to flattening out the frequency response.

## 6.7   Simulation examples

In this section, we consider three simulation examples to illustrate the features of the proposed procedure. For evaluating robustness, we use the peak of the sensitivity function

$$M_S \doteq \|S\|_\infty = \sup_\omega \left| \frac{1}{1 + PK(j\omega)} \right| \tag{6.32}$$

Because $M_S$ is the inverse of the shortest distance from the Nyquist curve of $L = PK$ to the critical point $-1 + 0j$, small values of $M_S$ indicate good robustness. For a reasonably robust system, an upper bound for the $M_S$ value can be fixed at around two [72]. Another robustness indicator used throughout the examples is given by $M_T \doteq \|T\|_\infty \doteq \sup_\omega |T(j\omega)|$.

The robustness interpretation for $M_T$ (the peak of $|T|$) comes from the robust stability condition (6.28). To quantify the input usage, we compute the total variation (TV) of the input $u$:

$$\text{TV} \approx \sum_{i=1}^{\infty} |u_{i+1} - u_i| \tag{6.33}$$

where $\{u_i\}_{i=1}^{\infty}$ denotes a discretization sequence of $u$. In the examples that follow, we restrict our attention to (unity) step disturbances ($d_d = s$) as it is commonly done in the literature.

**Example 1** The purpose of this preliminary example is to illustrate the different effect of the $\lambda$ and $\gamma$ parameters, completing the discussion in Section 6.3. We will consider the process $\tilde{P} = \frac{-100(10s+1)(0.02s+1)}{(-100s+1)(s+1)(0.2s+1)}$, modeled as $P = \frac{-100(10s+1)}{(-100s+1)(s+1)}$. The proposed design yields $Q = P_m^{-1}f = P^{-1}f$, where

$$f = \frac{a_1 s + 1}{(\lambda s + 1)(\gamma s + 1)} \tag{6.34}$$

with $a_1 = 100\left[(1 + \lambda/100)(1 + \gamma/100) - 1\right]$ and $\gamma \in [\lambda, 100]$. For $\gamma = \lambda$, (6.34) coincides with the conventional IMC filter used in [49, 43], which in this case favours input disturbances. As a consequence, the response for output disturbances may be undesirable. Figure 6.3 displays the time/frequency responses for $\lambda = \gamma = 0.15$. As can be seen, the peak in $|T|$ ($M_T = 1.16$) degrades robust stability, read as $|T(j\omega)| < 1/|\Delta(j\omega)| \ \forall \omega$, and it is responsible for the large oscillations in the response to the output disturbance. We know that by increasing $\gamma$ (we take $\gamma = 20$) it is possible to improve this response. As stated in Section 3, this tends to make the system slower too. In order to preserve the original closed-loop bandwidth, the $\lambda$ parameter can be decreased (we finally take $\lambda = 0.06$). As shown in Section 6.3, this retuning allows keeping the original closed-loop bandwidth while avoiding the peak in $|T|$ (now, $M_T = 1$). The resulting outcome is better robustness and smoother response. It is remarkable that it is not possible to press down the peak of $|T|$ by using the classical filter (for which $\gamma = \lambda$), as illustrated in Figures 6.4 and 6.5. Clearly, if one uses the standard filter structure, the only reasonable option is to detune the controller, moving the peak of $|T|$ to lower frequencies (see Figure 6.5). This will improve robustness at the expense of nominal performance. In summary, even if there is an interaction between $\lambda$ and $\gamma$, their roles are significantly different.

**Example 2** As pointed out in the early work [63], an optimal controller designed for a specific type of disturbance (e.g., a step acting at the input of the plant) may result

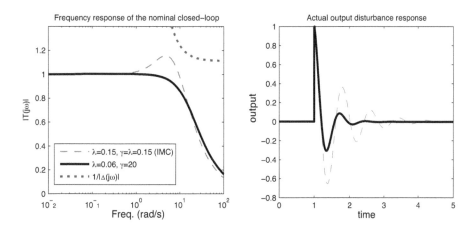

**FIGURE 6.3**
Frequency and time responses (Example 1), $\Delta = \frac{\tilde{P} - P}{P}$.

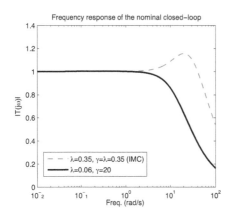

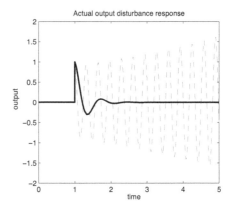

**FIGURE 6.4**
Frequency and time responses (Example 1).

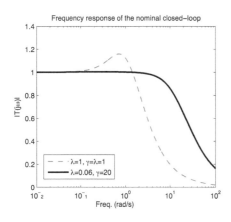

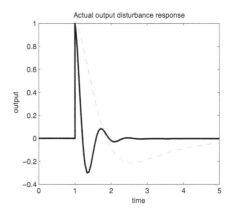

**FIGURE 6.5**
Frequency and time responses (Example 1).

in very poor performance if the actual disturbance (e.g., a step acting at the output) is different from the one considered at the design stage. In this example, we examine how a balance between the response of input and output disturbances can be achieved, focusing on the FOPTD model $P = K_g \frac{e^{-sh}}{\tau s + 1}$ ($K_g = 2, h = 1, \tau = 15$). The controller was already calculated in (6.14), and the corresponding filter has the same form as (6.34), taking now $a_1 = \tau - \tau(1 - \lambda/\tau)(1 - \gamma/\tau)$. Let us start by selecting $\lambda = \gamma = 1.75$, which provides $M_S = 1.69, M_T = 1.28, \text{TV}_i = 2.98, \text{TV}_o = 39.78$ ($\text{TV}_i, \text{TV}_o$ denote, respectively, the total variation with respect to the input and output disturbance). As shown by Figure 6.6(a), good attenuation of load disturbances is obtained. However, a somewhat large undershoot occurs for output disturbances. In the uncertain case ($h = 1.9$), we can see that the system becomes quite oscillatory ($\text{TV}_i = 10.72, \text{TV}_o = 82$); see Figure 6.6(b). By choosing $\gamma = \tau = 15$, we can avoid the undershoot in the output disturbance response, and improve the robustness margins, now $M_S = 1.33, M_T = 1, \text{TV}_i = 2$, and $\text{TV}_o = 16.14$. As a result, the responses are smoother in the uncertain case ($\text{TV}_i = 2.52, \text{TV}_o = 18.6$), experiencing less performance degradation. However, the performance for load disturbances is poor, showing a sluggish return to steady state (this fact, sometimes referred to as *loss of integral action*, is specially relevant for very lag-dominant plants with high gain [63]). To reach a compromise, we finally retune the controller taking $\lambda = 0.9, \gamma = 5$. The latter

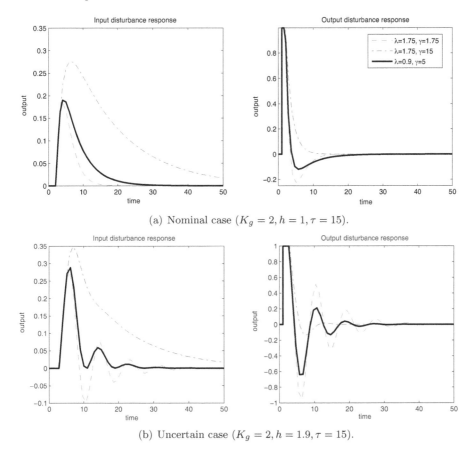

(a) Nominal case ($K_g = 2, h = 1, \tau = 15$).

(b) Uncertain case ($K_g = 2, h = 1.9, \tau = 15$).

**FIGURE 6.6**
Input/output step responses for Example 2.

values give $M_S = 1.6, M_T = 1.13$, and $\text{TV}_i = 2.55, \text{TV}_o = 40.67$. In the uncertain case, $\text{TV}_i = 6.73, \text{TV}_o = 67.63$. The data concerning this example has been collected in Table 6.1.

**Example 3** Lastly, we consider an stable second-order system with a pair of poorly damped poles. The model is given by $P = K_g \dfrac{e^{-sh}}{\left(\frac{s}{\omega_n}\right)^2 + 2\frac{\xi}{\omega_n}s + 1}$ ($K_g = 4, h = 1, \omega_n = 0.5, \xi = 0.25$). Our design suggests the controller $Q = P_m^{-1} f = \left((s/\omega_n)^2 + 2\xi/\omega_n s + 1\right) f$, where $f$ has the following structure

$$f = \frac{a_2 s^2 + a_1 s + a_0}{(\lambda s + 1)^2 \left(\gamma_{1,2} s^2 + \gamma_{1,1} s + 1\right)} \tag{6.35}$$

**TABLE 6.1**
Data summary for Example 2.

| Model: $K_g \frac{e^{-sh}}{\tau s+1}$ | | | Nominal Case $K_g = 2, h = 1, \tau = 15$ | | | | Uncertain Case $K_g = 2, h = 1.9, \tau = 15$ | | | |
|---|---|---|---|---|---|---|---|---|---|---|
| | | | Input dist. | | Output dist. | | Input dist. | | Output dist. | |
| Tuning of (6.34) | $M_S$ | $M_T$ | TV | ISE | TV | ISE | TV | ISE | TV | ISE |
| $\lambda = 1.75, \gamma = 1.75$ | 1.69 | 1.28 | 2.98 | 0.17 | 39.78 | 2.39 | 10.72 | 0.31 | 82 | 6.3 |
| $\lambda = 1.75, \gamma = 15$ | 1.33 | 1 | 2 | 0.66 | 16.14 | 2.34 | 2.52 | 0.69 | 18.6 | 3.06 |
| $\lambda = 0.9, \gamma = 5$ | 1.6 | 1.13 | 2.55 | 0.21 | 60.67 | 2.71 | 6.73 | 0.29 | 67.63 | 5.92 |

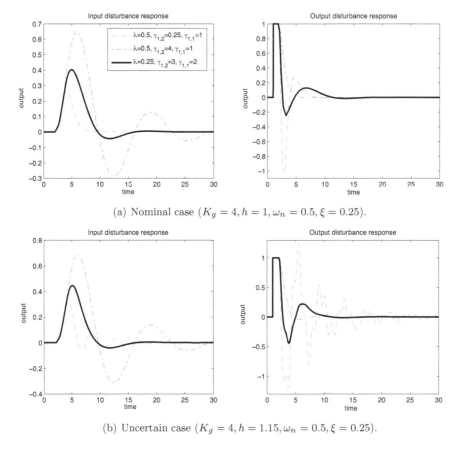

(a) Nominal case ($K_g = 4, h = 1, \omega_n = 0.5, \xi = 0.25$).

(b) Uncertain case ($K_g = 4, h = 1.15, \omega_n = 0.5, \xi = 0.25$).

**FIGURE 6.7**
Input/output step responses for Example 3.

and the $a_i$ coefficients satisfy that $P_a f = e^{-s} f = 1$ at the poles of $W = \frac{(\lambda s+1)^2(\gamma_{1,2}s^2+\gamma_{1,1}s+1)}{s((s/\omega_n)^2+2\xi/\omega_n s+1)}$. First, we select $\lambda = 0.5$. For output disturbances, we then take $\gamma_{1,2} = (1/\omega_n)^2 = 4, \gamma_{1,1} = 2\xi/\omega_n = 1$, which results into $f = \frac{1}{(0.5s+1)^2}$. The associated responses can be seen in Figure 6.7 for both the nominal and the uncertain cases. The tuning $\lambda = 0.5, \gamma_{1,2} = 4, \gamma_{1,1} = 1$ gives $M_S = 1.62, M_T = 1, \mathrm{TV}_i = 4.91, \mathrm{TV}_o = 140$ (nominal case) and $\mathrm{TV}_i = 6.5, \mathrm{TV}_o = 142$ (uncertain case). This design gives good results for output disturbances because the slightly damped poles are cancelled by the feedback controller. However, as no additional damping is really provided, these modes appear when excited from the input of the plant. Consequently, the response to load disturbances is quite oscillatory. To obtain much better performance for load disturbances, we select $\lambda = 0.5, \gamma_{1,2} = \lambda^2 = 0.25, \gamma_{1,1} = 2\lambda = 1$. For these settings, the filter is $f = \frac{2.9658s^2+2.3389s+1}{(0.5s+1)^3}$. As desired, the response to load disturbances has been improved noticeably. However, a great undershoot appears for output disturbances, indicating that robustness has been seriously degraded: $M_S = 3.88, M_T = 2.96$. The corresponding input usage is given by $\mathrm{TV}_i = 41.44, \mathrm{TV}_o = 1840$ (nominal case) and $\mathrm{TV}_i = 236.8, \mathrm{TV}_o = 8365$ (uncertain case, $h = 1.15$). A trade-off between the two designs considered so far can be obtained by selecting $\gamma = 0.25, \gamma_{1,2} = 3, \gamma_{1,1} = 2$, which corresponds to the filter $f = \frac{4.682s^2+2.06s+1}{(0.25 \text{ and } s+1)(3s^2+2s+1)}$. With this retuning, we finally get $M_S = 2.21, M_T = 1.35, \mathrm{TV}_i = 17.21, \mathrm{TV}_o = 903$

**TABLE 6.2**

Data summary for Example 3.

| Model: $K_g \dfrac{e^{-sh}}{\left(\frac{s}{\omega_n}\right)^2 + 2\frac{\xi}{\omega_n}s + 1}$ | | | Nominal Case $K_g = 4, h = 1, \omega_n = 0.5, \xi = 0.25$ | | | | Uncertain Case $K_g = 4, h = 1.15, \omega_n = 0.5, \xi = 0.25$ | | | |
|---|---|---|---|---|---|---|---|---|---|---|
| | | | Input dist. | | Output dist. | | Input dist. | | Output dist. | |
| Tuning of (6.35) | $M_S$ | $M_T$ | TV | ISE | TV | ISE | TV | ISE | TV | ISE |
| $\lambda = 0.5, \gamma_{12} = 0.25, \gamma_{11} = 1$ | 3.88 | 2.96 | 41.44 | 0.59 | 1840 | 11.55 | 236.8 | 0.85 | 8365 | 33.82 |
| $\lambda = 0.5, \gamma_{12} = 4, \gamma_{11} = 1$ | 1.62 | 1 | 4.91 | 3.78 | 140 | 8.19 | 6.5 | 4.73 | 142 | 8.34 |
| $\lambda = 0.25, \gamma_{12} = 3, \gamma_{11} = 2$ | 2.21 | 1.35 | 17.21 | 2.14 | 903 | 9.78 | 24.4 | 2.62 | 1121 | 12.5 |

(nominal case) and $\mathrm{TV}_i = 24.4, \mathrm{TV}_o = 1121$ (uncertain case). The idea for selecting $\gamma_{1,2}, \gamma_{1,1}$ is to place the complex poles of $f$ to the left of those of the plant $P$, and with increased damping factor. A summary of the results obtained can be consulted in Table 6.2.

To conclude this example, we will consider the simplified structure for $f$ given by (6.25), which in the case at hand has the form

$$f = \frac{a_2 s^2 + a_1 s + a_0}{(\lambda s + 1)^2 (\gamma s + 1)^2} \tag{6.36}$$

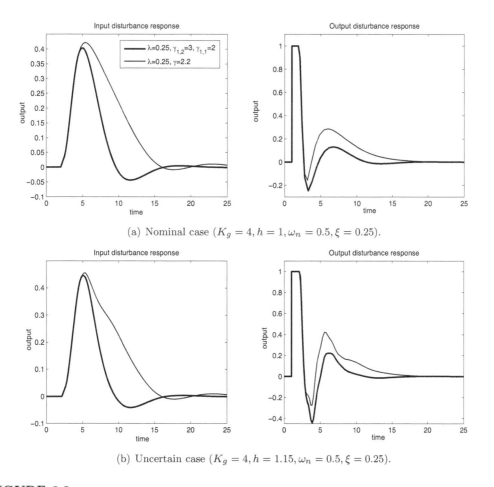

(a) Nominal case ($K_g = 4, h = 1, \omega_n = 0.5, \xi = 0.25$).

(b) Uncertain case ($K_g = 4, h = 1.15, \omega_n = 0.5, \xi = 0.25$).

**FIGURE 6.8**

Time responses for Example 3 with simplified filter structure.

where the $a_i$ coefficients satisfy that $P_a f = e^{-s} f = 1$ at the poles of $W = \frac{(\lambda s+1)^2 (\gamma s+1)^2}{s\left((s/\omega_n)^2 + 2\xi/\omega_n s+1\right)}$. This filter has the same structure suggested in [22] for MP unstable plants. If we choose $\gamma = \lambda$, we recover the design for input disturbances. The purpose now is to show that, although this filter can also be used to improve robustness with respect to the design for input disturbances, the robustness enhancement requires in general sacrificing more nominal performance than when using the full-structure filter with $\gamma_{1,2}, \gamma_{1,1}$. Taking $\lambda = 0.25, \gamma = 2.2$, the concrete filter $f = \frac{8.354s^2 + 3s + 1}{(0.25s+1)(2.2s+1)^2}$ results, for which $M_S = 2.24, M_T = 1.4$. These robustness indicators are only a little worse than those obtained for the previous trade-off tuning ($M_S = 2.21, M_T = 1.35$). However, the overall performance is considerably worse as it can be appreciated from Figure 6.8. This shows the necessity of considering complex conjugate poles in the filter $f$ for the best trade-off design.

# Part III

# Weighted Sensitivity Approach for Robust PID Tuning

# 7

## PID Design as a Weighted Sensitivity Problem

This chapter takes the results presented so far in the preceding chapters. The presentation is quite self-contained with the purpose of providing a compact perspective of the PID design problem within the weighted sensitivity problem but without explicit reference and *distractions* that may arise because of more formal and technical appreciations. In that respect, the chapter exposes first the inherent control design trade-off of a control system. This drives us in Section 7.3 to the generic PID controller design problem. The generic weights presented in the previous part are adapted here to the particular case of a PID design problem and the problem analytically solved for the generic case.

## 7.1  Context, motivation, and objective

At the core of process control we find an indispensable tool: the PID controller, whose ideal transfer function was introduced in Section 1.2. Although many changes and innovations have been introduced since its early development during the 1930s and '40s, the basic idea behind the PID controller still applies successfully in practice. Moreover, in spite of other promising proposals, such as the fuzzy or MPC paradigms [18, 56, 52], it seems that PID control is here to stay as the preferred control algorithm, at least at the bottom layer. This was the conclusion reached at the end of the IFAC Conference on Advances in PID Control, held in Brescia (Italy) during 28-30 March, 2012.

Even if only three parameters have to be tuned in a PID controller, it is not easy to find good settings without a systematic procedure [71, 31]. As a matter of fact, although tuning rules have been around almost as long as PID controllers, incorrect controller settings (in both manual and automatic control modes) still constitute a limiting factor for product-quality and process-operability [69, 37, 73]. This might be one of the reasons why PID tuning continues to be an active research field. Another reason for that is to better characterize the inherent design trade-offs by incorporating elements of modern theories, such as optimal and $\mathcal{H}_\infty$ control [55, 32, 81, 10], or by means of numerical optimization [19, 39, 12, 31, 29].

Among the existing tuning rules, the SIMC method [69] for PID controller tuning has become very popular. The key reasons for its success are mainly two: on the one hand, the tuning expressions are particularly simple (even easy to remember) and well-motivated by analytical derivations; on the other hand, they yield good performance[1]. These features make the SIMC method very suitable for training purposes in both academia and industry. Besides, the SIMC method allows adjustment of the robustness/performance trade-off by means of a single tuning parameter governing the closed-loop bandwidth, while at the same time guaranteeing a good balance between servo and regulatory control [31]. We want to emphasize the latter point because, in the literature, performance normally focuses on either

---

[1]In fact, it is not possible to do much better than SIMC with any other tuning, at least for PI control based on a (stable) first order with time-delay model [72, 31].

servo or regulatory objectives, see for example the concrete works [81, 12, 29], or [45] for a historical panorama.

Sometimes, however, we have both servo and regulatory requirements, for instance:

- In cascade configurations: the inner loop should be tuned based on tracking as it receives the set-points from the master loop. However, the inner loop may also need acceptable input disturbance supression capabilities [73].

- In cases with both input and output disturbances [75, 48, 71, 31], or when one simply doesn't know where the disturbance may occur [63]. Let us take a heat exchanger where steam is the heating medium, for example. Any of the following disturbances is possible: change in process flow rate, change in process inlet temperature, change in steam pressure, and change in specific heat (unlikely) [73].

In the above motivating examples, an intermediate tuning in between servo and regulation might be helpful to come up with an overall good solution. A limitation of the SIMC method is that only stable processes are considered [69, 66]. Indeed, tuning of stable and unstable plants is normally assessed separately, often employing different methodologies. See, for example, [93, 45, 50] and the references therein. In this regard, the work by [50], which focuses on regulatory performance only, constitutes one of the few ones unifying the treatment of stable and unstable plants.

With this context in mind, our basic aim is to give some new insights into the tuning problem by considering a unifying approach to take care of the most relevant conflicting objectives, namely the robustness/performance and servo/regulation trade-offs [45, Section 2.5]. The general idea is to obtain tuning expressions where the design parameters show how to shift each compromise. Towards this objective, we deal with stable and unstable plants analytically in a unified way, similarly as done in [50]. In a second phase (see Chapter 8), we concentrate on giving tuning guidelines with an emphasis on balanced servo/regulation operation along the lines of [69, 31]. The groundwork for achieving these goals was laid by the design methodology presented in Chapter 5, which meets all the requirements that we need.

## 7.2   Servo/regulation and robustness/performance trade-offs

In general, the tuning of the controller must be done taking into account different objectives, such as output performance, robustness, input usage, and noise sensitivity. Although, in principle, we should confine ourselves to multiobjective optimization, the trade-off space is recognized as having only one main dimension, namely high versus low controller gain [71, 31]. This inherent trade-off can be explained in terms of the sensitivity and complementary sensitivity functions, respectively:

$$S \doteq \frac{1}{1+PK} \qquad T \doteq 1-S = \frac{PK}{1+PK} \tag{7.1}$$

For performance we want $S \approx 0$ ($T \approx 1$), corresponding to high controller gain or *tight* control, whereas for robustness (also for noise sensitivity and input usage) we want $T \approx 0$ ($S \approx 1$), corresponding to low controller gain or *smooth* control [72, 70]. Since

$$S + T = 1, \tag{7.2}$$

it is clear that one cannot improve robustness and peformance at the same time, implying that a *good* balance over frequency has to be found in practice. This is the well-known trade-off between speed of response (i.e., closed-loop bandwidth) and robustness [49, 72].

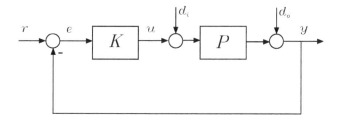

**FIGURE 7.1**
Unity feedback scheme. Output disturbances ($d_o$) can be regarded as unmeasured set-point changes ($r = -d_o$).

From the point of view of ouput performance, we can identify another design trade-off by looking at the closed-loop effect of disturbance and set-point changes on the output error in Figure 7.1, which is given by

$$e = r - y = S(r - d_o) + S_i d_i, \tag{7.3}$$

where

$$S_i = SP \tag{7.4}$$

As explained in [48], if the plant $P$ has slow[2] stable poles, there is an inevitable trade-off between having the slow, stable open-loop poles appearing as zeros of $S$, producing overshoot in the closed-loop responses to set-point and output disturbance changes, or as poles of $S_i$, producing a sluggish response to input disturbances. Since the limitations imposed by slow stable poles are similar (although less severe) to those imposed by slow unstable ones, the same kind of trade-off between $S$ and $S_i$ appears when considering unstable poles. Thus, in the extreme cases, the controller $K$ of Figure 7.1 may generally be tuned either in *regulator* mode, where the input disturbance $d_i$ is the main exogenous input and the design typically aims at minimizing $\|\frac{1}{s}S_i\|$ (in some usual norm), or in servo mode, where the set-point change $r$ and the output disturbance $d_o$ are the main concern and the design seeks to minimize $\|\frac{1}{s}S\|$ [49, 39]. Because the reference signal $r$ can be shaped using set-point weighting [15, 11], it is possibly more interesting to study the here-called servo/regulation trade-off with respect to transients caused by unmeasured input/output disturbances[3] [48, 71]. Also, because designs aimed at load disturbance rejection exhibit larger peaks on $|T|$, there are yet other ways of looking at the considered compromise, for example, in terms of input disturbance rejection versus sensitivity to modelling errors [22, 48]. At the end of the day, what really matters is that Equation (7.4) has a fundamental impact, similarly as the better-known complementary constraint of Equation (7.2) [48].

## 7.3    Unifying tuning rules

Let us consider the single-input single-output (SISO) feedback setup of Figure 7.1, where $P$ and $K$ represent the plant and the controller, and $r, d_i, d_o, u,$ and $y$ denote the reference (or set-point), an input (or load) disturbance, an output disturbance, the

---

[2]By an slow open-loop pole, we mean one below the desired closed-loop bandwidth.

[3]In this regard, other terms for the trade-off, as for example input/output disturbance, $S/SP$, or $S/S_i$, would probably be more appropriate. The term servo/regulation has been preferred for consistency with previous works of the authors and with the nomenclature commonly used within the PID literature [12, 10, 71].

control and the output signals, respectively. In particular, in this chapter, we will focus on PID control, and the following ideal form for the controller will be assumed[4]:

$$K = K_c \left(1 + \frac{1}{T_i s} + T_d s\right) \tag{7.5}$$

where $K_c$ is the controller gain, $T_i$ the integral time and $T_d$ the derivative time. For any industrial plant without resonant characteristics, the dominant dynamics can be represented by the second-order model[5]

$$P = \frac{k e^{-\theta s}}{(\tau_1 s + 1)(\tau_2 s + 1)} \tag{7.6}$$

where $k$ is the process gain, $\tau_1$ the dominant lag time constant, $\tau_2$ is the second-order lag time constant, and $\theta$ the (effective) time delay. To account for unstable plants, $\tau_1$ and/or $\tau_2$ may be negative. Finally, the exogenous signals $r, d_i, d_o$ will be assumed to be step-like signals as is customary in the literature.

For design purposes, we now rely on the well-known WSP [72, Section 2.8.2]:

$$\min_{K \in \mathcal{C}} \|WS\|_\infty \tag{7.7}$$

where $\mathcal{C}$ denotes the set of internally stabilizing controllers, $\|WS\|_\infty \doteq \max_\omega |W(j\omega)S(j\omega)|$ (the peak of the magnitude frequency response), and $W$ is a weight responsible for the shaping of the sensitivity function $S$. More concretely, the following structure for the weight is adopted from the preceding chapters:

$$W = \frac{(\lambda s + 1)(\gamma_1 s + 1)(\gamma_2 s + 1)}{s(\tau_1 s + 1)(\tau_2 s + 1)} \tag{7.8}$$

where

$$\lambda > 0, \gamma_i \in [\lambda, |\tau_i|], i = 1, 2. \tag{7.9}$$

The integrator $1/s$ is included in $W$ to provide integral action to the loop. The rationale behind the $\lambda$ and $\gamma$ parameters can be explained, quite heuristically, as follows. Start by considering $\lambda \approx 0$, then:

- If $\gamma_i = |\tau_i|, i = 1, 2$, the weight $W$ looks like $\frac{1}{s}$, as a matter of fact $|W| \approx |\frac{1}{s}|$, and we have that the optimization problem (7.7) minimizes $\|\frac{1}{s}S\|_\infty$ (subject to internal stability). As commented, this corresponds to a *servo* specification.

- If $\gamma_i = \lambda \approx 0, i = 1, 2$, the weight $W$ looks like $\frac{1}{s}P$ because we have included the poles of $P$ in $W$, in particular $|W| \approx |\frac{1}{s}P|$ (we assume $k = 1$ in (7.6) without loss of generality), and accordingly the optimization problem (7.7) basically minimizes $\|\frac{1}{s}S_i\|_\infty$. This corresponds to a *regulator* specification.

Obviously, intermediate values for $\gamma_1$ and $\gamma_2$ will produce a balance between the purely servo and regulation situations. Finally, if we now increase the value of $\lambda$ (so far we had assumed $\lambda \approx 0$), the weight will progressively slow down the resulting closed loop. Thus, once $\gamma_1, \gamma_2$ have been fixed, $\lambda$ can be used to reach a balance between robustness and performance. In summary, the selected weight allows us to deal with both robustness/performance (via $\lambda$) and servo/regulation issues (via $\gamma_1, \gamma_2$).

---

[4]For simulation purposes, however, the commercial PID form $K = K_c \left(1 + \frac{1}{T_i s} + \frac{T_d s}{\alpha T_d s + 1}\right)$ is used. In order to not bias the results with respect to the ideal (non-realizable) PID controller (7.5), we select $\alpha = 0.001$.

[5]The modelling issue will not be covered here. For details on how to obtain the model, we refer the reader to [66], where both open-loop and closed-loop step-response tests are presented. Also, when a high-order description of the plant is the starting point, one can reduce the order of the model using the *half-rule* method [69].

**Remark 7.3.1.** *It is noteworthy that the weight (7.8) is unstable when $\tau_1 < 0$ or $\tau_2 < 0$, that is, when the plant itself is unstable. Unstable weights are not common in $\mathcal{H}_\infty$ control. However, as it was shown in Chapter 5, including the (possibly unstable) poles of the plant in $W$ constitutes a plausible trick to avoid coprime factorizations, henceforth unifying the treatment of stable and unstable plants.*

For analytical purposes, we first need to approximate the time delay in the SOPTD model (7.6). For simplicity, we take $e^{-\theta s} \approx -\theta s + 1$, resulting into the following model approximation:

$$P \approx \frac{k(-\theta s + 1)}{(\tau_1 s + 1)(\tau_2 + 1)} \tag{7.10}$$

Now, from Theorem 5.2.1 (or alternatively, by application of the maximum modulus principle [72, page 173]), it turns out that the optimal weighted sensitivity in (7.7) is constant[6] over frequency, that is, we have that $WS^o = \rho$. Moreover, $\rho$ is given by

$$\rho = |W|_{s=\frac{1}{\theta}} = \frac{(\lambda + \theta)(\gamma_1 + \theta)(\gamma_2 + \theta)}{(\tau_1 + \theta)(\tau_2 + \theta)} \tag{7.11}$$

Straightforward algebra leads us to the optimal controller

$$K^o = P^{-1}(\rho^{-1}W - 1) = \frac{\chi}{\rho k s} = \frac{\zeta_2 s^2 + \zeta_1 s + 1}{\rho k s} \tag{7.12}$$

where

$$\zeta_1 = \frac{\theta\left((\tau_1 + \tau_2 - \lambda)(\gamma_1 + \gamma_2) + \lambda(\tau_1 + \tau_2)\right) + \tau_1\tau_2(\gamma_1 + \gamma_2 + \lambda + \theta)}{-\gamma_1\gamma_2(\lambda + \theta) + \theta^2(\tau_1 + \tau_2)}{(\tau_1 + \theta)(\tau_2 + \theta)} \tag{7.13}$$

$$\zeta_2 = \frac{\tau_1\tau_2\left((\lambda + \theta)(\gamma_1 + \gamma_2) + \gamma_1\gamma_2 + \lambda\theta + \theta^2\right) - \gamma_1\gamma_2\lambda(\theta + \tau_1 + \tau_2)}{(\tau_1 + \theta)(\tau_2 + \theta)} \tag{7.14}$$

Comparing (7.12) with (7.5), we readily observe that $K^o$ is, in fact, a PID compensator tuned as follows:

$$K_c = \frac{\zeta_1}{\rho k} \quad T_i = \zeta_1 \quad T_d = \frac{\zeta_2}{\zeta_1} \tag{7.15}$$

---

## 7.4 Special cases and tuning-rule simplifications

There are some special cases of (7.6) which deserve specific attention. We will now detail the tuning rules for these specific cases. In addition, to simplify the tuning expressions (7.13)–(7.15) (and the tuning process itself), we propose to using only one $\gamma$ parameter, instead of $\gamma_1$ and $\gamma_2$.

### 7.4.1 First-order cases ($\tau_2 = 0$)

Most controller tuning rules require only three pieces of information: process gain $k$, dead time $\theta$, and process lag or time constant $\tau_1$ [73]. The corresponding step-response model in those cases is:

$$P = \frac{ke^{-\theta s}}{\tau_1 s + 1} \tag{7.16}$$

---

[6]More generally, the characterization of the optimal (weighted) sensitivity as an all-pass function is due to [90].

Note that the FOPTD model (7.16) is a special case of (7.6) when $\tau_2 = 0$. For such a process description, we can take $\gamma = \gamma_1, \gamma_2 = 0$ and the following PI settings ($T_d = 0$) result from (7.15):

$$
\begin{aligned}
K_c &= \frac{\tau_1(\gamma + \lambda + \theta) - \lambda\gamma}{k(\lambda + \theta)(\gamma + \theta)} \\
T_i &= \frac{\tau_1(\gamma + \lambda + \theta) - \lambda\gamma}{\tau_1 + \theta}
\end{aligned}
\tag{7.17}
$$

where $\lambda > 0, \gamma \in [\lambda, |\tau_1|]$. An important particular case of (7.16) arises by examining considerably lag-dominant plants ($|\tau_1| \gg \theta$). Then we have that

$$
P = \frac{ke^{-\theta s}}{\tau_1 s + 1} = \frac{k'e^{-\theta s}}{(s + 1/\tau_1)} \approx \frac{k'}{s}e^{-\theta s}
\tag{7.18}
$$

where $k' = \frac{k}{\tau_1}$ is the so-called integration rate or velocity constant. The previous approximation is exact when $|\tau_1| \to \infty$, corresponding to an IPTD process. In this case, the tuning rule (7.17) simplifies to:

$$
\begin{aligned}
K_c &= \frac{\gamma + \lambda + \theta}{k'(\lambda + \theta)(\gamma + \theta)} \\
T_i &= \gamma + \lambda + \theta
\end{aligned}
\tag{7.19}
$$

where $\lambda > 0, \gamma \in [\lambda, \infty)$. In concrete, the proportional controller $K_c = \frac{1}{k'(\lambda+\theta)}$ is obtained as we make $\gamma \to \infty$ in (7.19). This controller is excellent for servo control, but results in steady-state offset for input disturbances. Clearly, it is for the integrating case that we have the most stringent trade-off between servo and regulatory control (this fact is also indicated by the range of variation of the $\gamma$ parameter).

### 7.4.2   Second-order cases

A nice feature of the PI settings (7.17) and (7.19) is that only one parameter ($\gamma$) is used to adjust the servo/regulation trade-off, in contrast with (7.15), where two tuning parameters ($\gamma_1, \gamma_2$) are used for the same purpose. In order to simplify (7.15), we can set $\gamma_1 = \gamma_2 = \gamma$, which finally yields

$$
\begin{aligned}
K_c &= \frac{\theta\left(2\gamma(\tau_1 + \tau_2 - \lambda) + \lambda(\tau_1 + \tau_2)\right) + \tau_1\tau_2(2\gamma + \lambda + \theta) - \gamma^2(\lambda + \theta) + \theta^2(\tau_1 + \tau_2)}{k(\lambda + \theta)(\gamma + \theta)^2} \\
T_i &= \frac{\theta\left(2\gamma(\tau_1 + \tau_2 - \lambda) + \lambda(\tau_1 + \tau_2)\right) + \tau_1\tau_2(2\gamma + \lambda + \theta) - \gamma^2(\lambda + \theta) + \theta^2(\tau_1 + \tau_2)}{(\tau_1 + \theta)(\tau_2 + \theta)}
\end{aligned}
\tag{7.20}
$$

$$
T_d = \frac{\tau_1\tau_2\left(2\gamma(\lambda + \theta) + \gamma^2 + \lambda\theta + \theta^2\right) - \gamma^2\lambda(\theta + \tau_1 + \tau_2)}{\theta\left(2\gamma(\tau_1 + \tau_2 - \lambda) + \lambda(\tau_1 + \tau_2)\right) + \tau_1\tau_2(2\gamma + \lambda + \theta) - \gamma^2(\lambda + \theta) + \theta^2(\tau_1 + \tau_2)}
$$

with $\lambda > 0$ and $\gamma \in [\lambda, \tau]$, where we have defined

$$
\tau \doteq \max(|\tau_1|, |\tau_2|) = |\tau_1|.
\tag{7.21}
$$

The simplification $\gamma = \gamma_1 = \gamma_2$ is particularly well motivated when $\tau_1 = \tau_2$, that is, when (7.6) has a double real pole. The tuning expressions for this case are given below:

$$
K_c = \frac{2\theta\left(\gamma(2\tau_1 - \lambda) + \lambda\tau_1\right) + \tau_1^2(2\gamma + \lambda + \theta) - \gamma^2(\lambda + \theta) + 2\theta^2\tau_1}{k(\lambda + \theta)(\gamma + \theta)^2}
$$

$$T_i = \frac{2\theta \left(\gamma(2\tau_1 - \lambda) + \lambda\tau_1\right) + \tau_1^2(2\gamma + \lambda + \theta) - \gamma^2(\lambda + \theta) + 2\theta^2\tau_1}{(\tau_1 + \theta)^2} \tag{7.22}$$

$$T_d = \frac{\tau_1^2 \left(2\gamma(\lambda + \theta) + \gamma^2 + \lambda\theta + \theta^2\right) - \gamma^2\lambda(\theta + 2\tau_1)}{2\theta \left(\gamma(2\tau_1 - \lambda) + \lambda\tau_1\right) + \tau_1^2(2\gamma + \lambda + \theta) - \gamma^2(\lambda + \theta) + 2\theta^2\tau_1}$$

Let us now consider that $|\tau_1| \gg \theta$ in (7.6), then we have that:

$$P = \frac{ke^{-\theta s}}{(\tau_1 s + 1)(\tau_2 s + 1)} = \frac{k'e^{-\theta s}}{(s + 1/\tau_1)(\tau_2 s + 1)} \approx \frac{k'e^{-\theta s}}{s(\tau_2 s + 1)} \tag{7.23}$$

where $k' = \frac{k}{\tau_1}$. Thus, by making $|\tau_1| \to \infty$, we can particularize the settings (7.20) for the integrating with lag process, or second-order integrating plus time delay (SOIPTD) process, resulting into

$$
\begin{aligned}
K_c &= \frac{(\tau_2 + \theta)(\theta + 2\gamma + \lambda)}{k'(\lambda + \theta)(\gamma + \theta)^2} \\
T_i &= \theta + 2\gamma + \lambda \\
T_d &= \frac{\tau_2 \left((\gamma + \theta)^2 + \lambda(2\gamma + \theta)\right) - \gamma^2\lambda}{(\tau_2 + \theta)(\theta + 2\gamma + \lambda)}
\end{aligned}
\tag{7.24}
$$

where $\lambda > 0, \gamma \in [\lambda, \infty)$. Similarly, if, in addition, $|\tau_2| \gg \theta$, then

$$P \approx \frac{k'e^{-\theta s}}{s(\tau_2 s + 1)} = \frac{k''e^{-\theta s}}{s(s + 1/\tau_2)} \approx k'' \frac{e^{-\theta s}}{s^2} \tag{7.25}$$

where $k'' = \frac{k'}{\tau_2}$. Therefore, by making $|\tau_2| \to \infty$ in (7.24), or equivalently $|\tau_1| \to \infty$ in (7.22), we get the rules for the Second Order Double Integrating Plus Time Delay (SODIPTD) process:

$$
\begin{aligned}
K_c &= \frac{\theta + 2\gamma + \lambda}{k''(\lambda + \theta)(\gamma + \theta)^2} \\
T_i &= \theta + 2\gamma + \lambda \\
T_d &= \frac{(\gamma + \theta)^2 + \lambda(2\gamma + \theta)}{\theta + 2\gamma + \lambda}
\end{aligned}
\tag{7.26}
$$

where $\lambda > 0, \gamma \in [\lambda, \infty)$. It is easy to observe that, if we make $\gamma \to \infty$ in (7.26), we end up with a purely derivative controller for which the derivative gain is $K_D = K_c T_d = \frac{1}{k''(\lambda + \theta)}$. A large value of $\gamma$ gives nice results for servo control, but, as it can be easily verified, the resulting response to an input disturbance becomes a ramp. Again, it is for the (double) integrating case that we find the most severe servo/regulation trade-off.

## 7.5 Applicability: normalized dead time range

The multiple PID settings of Section 7.4 depend on the design parameters $\lambda, \gamma$, and on the plant model information $(k, \tau_1, \tau_2, \theta)$. Before embarking on the tuning of $\lambda$ and $\gamma$, we note the following facts: for stable plants, derivative action is only recommended for dominant second-order processes [69], i.e. $\tau_1 \geq \tau_2 > \theta$. Moreover, for plants dominated by the time delay ($\theta/\tau_2 \geq \theta/\tau_1 > 1$), there is only a benign trade-off between servo and regulatory responses [67, 2, 3, 31], and in such cases, one can simply select $\gamma = \tau_1$. On the other hand, in the unstable plant case, let us assume that $\tau_1 < 0$, and we have the following fundamental lower bound on the peak of $|T|$ [72, Theorem 5.4, page 173]:

$$M_T \doteq \|T\|_\infty \geq e^{\frac{\theta}{|\tau_1|}} \tag{7.27}$$

Thus, if $\theta/|\tau_1| = 1$, $M_T \geq e \approx 2.72$, whereas $M_T \geq e^{1.5} \approx 4.48$ for $\theta/|\tau_1| = 1.5$, and so on. In addition, from (7.2), it follows that $M_S \doteq \|S\|_\infty \geq M_T - 1$, and, consequently, large values of $M_T$ imply large values of $M_S$ too. Thus, in practice, $\theta/|\tau_1|$ should not be much larger than one, otherwise the unavoidable large values of $M_T$ and $M_S$ would indicate poor performance as well as robustness problems. These robustness issues are confirmed in the PID literature, where tuning rules aimed at unstable plants restrict themselves to small ranges of the normalized dead time [59, 50].

Based on the previous considerations concerning both stable and unstable processes, we will focus on lag-dominant ones, and, in order to make the study as symmetrical as possible with respect to stable/unstable plants, we mostly assume that

$$\frac{\theta}{|\tau_1|} \leq \frac{\theta}{|\tau_2|} < 1 \tag{7.28}$$

with an emphasis on near-integrating processes. If $\tau_2 = 0$, then we will simply take

$$\frac{\theta}{|\tau_1|} < 1 \tag{7.29}$$

# 8

## PID Tuning Guidelines for Balanced Operation

A side result of the general PID-tuning expressions in the previous chapter was the change of dimensionality with respect to the tuning of the PID controller. From the PID definition, the tuning problem is a three-dimensional search problem. With the tuning provided so far the problem is a two-dimensional one. In addition to this dimensionality reduction, there is an additional advantage of the provided PID tuning with respect to the meaning of the parameters. Now, with the $\gamma$ and $\lambda$, the parameters are related to the servo/regulation and robustness issues. How to balance them is the subject of this chapter.

In this chapter, we provide tuning guidelines for the $\lambda, \gamma$ parameters using the performance indices $J_{\max}, J_{\text{avg}}$ (or their global counterparts $J^*_{\max}, J^*_{\text{avg}}$) and the optimization problem (8.6) introduced in the previous chapter. The overall objective is to achieve a *balanced* closed-loop, that is, a good balance between servo and regulatory performance, on the one hand, and between robustness and performance, on the other hand.

### 8.1 Robustness and comparable servo/regulation designs

This section is devoted to explaining how to evaluate robustness and performance in order to assist in the tuning of $\lambda$ and $\gamma$. Before doing so, some space is devoted to understand the interplay between them, since, as one might guess, their roles are not perfectly decoupled. This understanding will be exploited in Sections 8.3 and 8.4 for tuning purposes. To see how $\lambda$ and $\gamma$ interact, let us consider Figure 8.1(a).

As it can be seen, for a given value of $\lambda$, servo and regulatory designs may have very different properties. More specifically, the regulatory design ($\gamma = \lambda$) is considerably less robust, due to the larger closed-loop bandwidth and sensitivity peaks, and faster, as it can be noticed from the shorter rise time in the set-point response. This result is not peculiar to our design example, but fundamental [48, Section 4]: in general, regulatory control is based on shifting the slow poles of the plant, whereas servo control aims at cancelling them (as long as possible) to flatten out the frequency response. When a slow pole of the plant is shifted by feedback, a zero slower than all the closed-loop poles appears in $T$. This zero, which necessarily raises $M_T$ above one, affects the closed-loop bandwidth and contributes to speed up the response time. Therefore, the comparison between the regulator and servo designs is left with a faster and less robust, versus a slower and more robust, alternative. Due to the big differences in the sensitivity peaks and closed-loop bandwidth, one cannot compare these alternatives directly.

To allow a meaningful comparison, [48] introduced the notion of *extreme frequency equivalence* to characterize transfer functions having the same relative degree, steady-state gain, and high-frequency behaviour. In particular, extreme frequency equivalent complementary sensitivities posses similar initial rise time and the same sensitivity to high-frequency noise and modelling errors. With this idea in mind, in Figure 8.1(b) we have increased the value of $\lambda$ in the regulator mode until making the servo and regulatory designs *comparable* in the

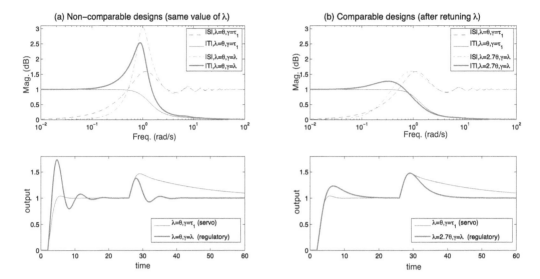

**FIGURE 8.1**

Influence of $\lambda$ and $\gamma$ parameters on closed-loop properties for $P = \frac{5e^{-s}}{20s+1}$ ($k = 5, \tau = \tau_1 = 20, \theta = 1$): frequency domain (top) and time domain responses for a unity set-point ($r$) change and load disturbance ($d_i$) at $t = 1$ and $t = 25$, respectively (bottom). Servo and Regulatory designs are made *comparable* in b) by imposing equal $M_S$ values. By doing so, similar rise time and closed-loop bandwidth are obtained.

extreme frequency equivalence sense. In general, we have observed that, by imposing the same value of $M_S$ ($\approx 1.6$ in the example), approximately the same closed-loop bandwidth and rise time are achieved regardless the value of $\gamma$. As a result, Figure 8.1b highlights that the trade-off between servo ($\gamma = \tau$) and regulatory control ($\gamma = \lambda$) is basically a matter of adjusting the peak on $|T|$, that is $M_T$. In this regard, regulation demands larger values of $M_T$, which renders regulatory designs more sensible to modelling errors (particularly near the crossover frequency) than the servo counterparts [48].

## 8.2 Servo/regulation performance evaluation: $J_{\mathbf{max}}$ and $J_{\mathbf{avg}}$ indices

From this point on, we will quantify robustness in terms of the sensitivity peak $M_S$. This decision has a two-fold justification. First, $M_S$ is commonly used as a robustness indicator because of its clear interpretation[1]: it gives the inverse of the distance between the Nyquist curve and the critical point $-1$ [29, 12]. Second, we have observed in Section 8.1 that fixing $M_S$ provides comparable designs for different adjustments of the servo/regulation trade-off, i.e., for different values of $\gamma$. The idea is that, if we reduce $\gamma$ for regulatory improvement, $\lambda$ must be increased to compensate for the robustness loss as illustrated in Figure 8.2.

Therefore, for each robustness level we can define

$$\Lambda\Gamma_k \doteq \left\{ (\lambda, \gamma); M_S = k, \lambda > 0, \gamma \in [\lambda, \tau] \right\} \tag{8.1}$$

---

[1]A less obvious advantage of $M_S$ is that it is highly correlated with input usage [71, 31].

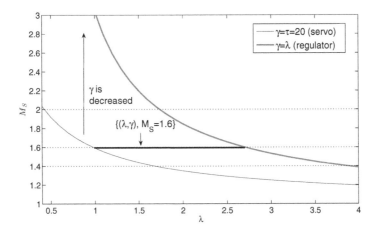

**FIGURE 8.2**

$M_S$ over $\lambda$ for extreme values of the $\gamma$ parameter and $P = \frac{5e^{-s}}{20s+1}$. If $\gamma$ is decreased to improve the regulatory performance, $\lambda$ must be increased accordingly to keep $M_S$ constant.

A problem to be considered is how to select $\gamma$ to yield a good balance between servo and regulatory performance. For this purpose, we consider the minimization of two alternative performance indices:

$$J_{\mathrm{max}} = \max(\Delta_s, \Delta_r) \tag{8.2}$$
$$J_{\mathrm{avg}} = 0.5(\Delta_s + \Delta_r) \tag{8.3}$$

where

$$\Delta_s = \frac{\mathrm{IAE}_s}{\mathrm{IAE}_s^o}, \qquad \Delta_r = \frac{\mathrm{IAE}_r}{\mathrm{IAE}_r^o} \tag{8.4}$$

and

$$\mathrm{IAE} = \int_0^\infty |r(t) - y(t)|dt = \int_0^\infty |e(t)|dt \tag{8.5}$$

In (8.4), the subindex $_s(_r)$ stands for servo(regulator), and indicates that the IAE is calculated over $\Lambda\Gamma_k$ with respect to set-point(load disturbance)[2]. Finally, the superindex $^o$ is used to indicate the optimal IAE value along with $\Lambda\Gamma_k$. The cost (8.3) is used in [71, 31] to evaluate the SIMC-PI method, and it weighs the importance of servo and regulatory performance about equally. As for the cost (8.2), already considered in [3], it adheres to the common strategy in multiobjective optimization of minimizing the worst-case performance. Importantly, both (8.2) and (8.3) are sound performance measures, independent of the process gain, the disturbance and set-point magnitudes, and of the units used for time. Ideally, we want $J_{\mathrm{max}} = J_{\mathrm{avg}} = 1$; this is the case where both optimal servo and regulatory responses are achieved. In practice, however, $J_{\mathrm{max}}, J_{\mathrm{avg}} > 1$ because of the inherent trade-off. Then, for each robustness level ($M_S = k$), we can consider the following optimization problem

$$\min_{(\lambda,\gamma)\in\Lambda\Gamma_k} J \tag{8.6}$$

---

[2]Since output disturbances $d_o$ can be treated as special set-point changes $r$, we will only consider set-point changes $r$ on the servo side from now on.

where $J = J_{\max}$ or $J = J_{\text{avg}}$. Note that in (8.4), performance degradation is measured with respect to the optimal IAEs restricted to $\Lambda\Gamma_k$ ($M_S = k$). This reference works well to compare servo and regulatory designs subject to a given robustness specification.

However, as the design gets more robust, $\lambda$ increases and the interval for $\gamma \in [\lambda, \tau]$ gets smaller (this can be appreciated from Figure 8.2), indicating that the servo/regulation trade-off tends to dissapear in high robustness designs, for which close-to-one values of $J_{\max}$ and $J_{\text{avg}}$ are obtained. This is misleading from a robustness/performance point of view, since high robustness should imply low performance, i.e., $J \gg 1$. Thus, in order to study the trade-off between robustness and performance in absolute terms, it will be better to consider the globally optimal IAE values in (8.4), that is, the optimal values over $\bigcup_k \Lambda\Gamma_k$, for $k$ ranging over all possible robustness levels. In this situation, we redefine performance degradation as follows

$$\Delta_s^* = \frac{\text{IAE}_s}{\text{IAE}_s^{go}}, \qquad \Delta_r^* = \frac{\text{IAE}_r}{\text{IAE}_r^{go}} \tag{8.7}$$

where the superindex $^{go}$ stands for *globally optimal* and $\text{IAE}_s, \text{IAE}_r$ are computed over $\Lambda\Gamma_k$ as before, for a given fixed value of $k$. In accordance, the performance indices (8.2) and (8.3) will be denoted by $J_{\max}^*$ and $J_{\text{avg}}^*$ when performance degradation is taken as in (8.7).

## 8.3  PI control using first-order models

We here provide autotuning guidelines for the PI settings (7.17) and (7.19), aimed at FOPTD and IPTD models, respectively. We first consider the stable ($\tau_1 > 0$) and integrating ($\tau_1 \to \infty$) plant cases, and, after that, we apply the same methodology to the unstable case ($\tau_1 < 0$).

### 8.3.1  Stable and integrating cases

Let us start by considering the trade-off between robustness and (servo/regulation) performance for different values of the normalized dead time. This trade-off is depicted in Figure 8.3. Although we are mainly interested in the combined performance indices $J_{\max}^*$ and $J_{\text{avg}}^*$, we have also included the corresponding purely servo ($\Delta_s^*$) and regulatory ($\Delta_r^*$) indices.

As for the different plotted curves ($J_{\max}^*, J_{\text{avg}}^*, \Delta_s^*, \Delta_r^*$), the first thing to note is that we are only interested in the region with negative or zero slope, corresponding to the left region of the graphs in Figure 8.3, approximately covering the range $M_S \leq 2$. Otherwise, we don't have a real trade-off between performance and robustness, and both of them can be improved simultaneously. Another important observation is that the gap between $\Delta_s^*$ and $\Delta_r^*$ is small for $\theta/\tau_1 = 1$, but increases as we consider smaller $\theta/\tau_1$ ratios, reaching its maximum in the integrating case ($\theta/\tau_1 = 0$). This indicates that the servo/regulation trade-off is important just for lag-dominanted processes, and especially for integrating ones. As already commented, this was predictable by looking at the interval of the $\gamma$ parameter in the tuning rule (7.17): $\gamma \in [\lambda, \tau_1]$, which grows with $\tau_1$ and it indeed becomes infinite for the IPTD process, i.e., when $\tau_1 = \infty$. Figure 8.3 also reveals that servo designs are inherently more robust than regulatory ones; while $\Delta_s^*$ reaches its minimum at around $M_S = 1.75$, $\Delta_r^*$ does so at $M_S \approx 2.15$ for processes with balanced lead-lag ratio, and at around $M_S = 3$ for $\theta/\tau_1 < 1$. So far we have looked at the purely servo and regulator designs. However, our objective is to provide guidelines for a well-balanced intermediate tuning. Because of this, we now turn our attention to the combined performance costs $J_{\max}^*$ and $J_{\text{avg}}^*$. As it is

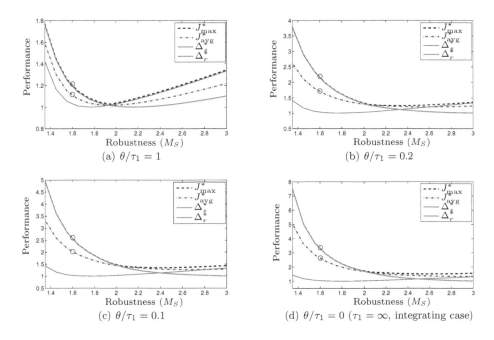

**FIGURE 8.3**
Robustness versus performance for FOPTD systems (pareto-optimal curves in the $\lambda\gamma$-space).

shown in Figure 8.3, they have a similar shape in the interesting region of the graphs. In general, for both the $J_{\max}^*$ and $J_{\text{avg}}^*$ criteria, a target robustness level of $M_S^t \approx 1.6$ gives a good trade-off between robustness and performance, corresponding to the circles located in the desired negative-slope regions of the graphs, quite to the left of the minima. According to this fact, we will select $M_S^t = 1.6$ as our target robustness level, and we will look at the solution of problem (8.6) for $k = 1.6$ and $J = J_{\max}$ and $J = J_{\text{avg}}$. The results have been collected in Table 8.1.

### 8.3.1.1  Tuning based on $J_{\max}$

Concerning the case $J = J_{\max}$, it has been found that the data for the column $\gamma/\lambda$ in Table 8.1 can be very well fitted by choosing $\gamma/\lambda = \frac{0.2\theta/\tau_1 + 0.9}{\theta/\tau_1 + 0.1}$. In practice, since $\gamma \in [\lambda, \tau_1]$, one should select $\gamma$ as follows

$$\gamma = \min\left(\frac{0.2\frac{\theta}{\tau_1} + 0.9}{\frac{\theta}{\tau_1} + 0.1}\lambda, \tau_1\right) \tag{8.8}$$

**TABLE 8.1**
Optimal values of $\lambda, \gamma$ for problem (8.6) taking $k = 1.6$ and a stable FOPTD system.

| $J = J_{\max}$ | | | | $J = J_{\text{avg}}$ | | | |
|---|---|---|---|---|---|---|---|
| $\theta/\tau_1$ | $\lambda/\theta$ | $\gamma/\theta$ | $\gamma/\lambda$ | $\theta/\tau_1$ | $\lambda/\theta$ | $\gamma/\theta$ | $\gamma/\lambda$ |
| 1 | 1 | 1 | 1 | 1 | 1 | 1 | 1 |
| 0.2 | 1.1887 | 3.7 | 3.1125 | 0.2 | 2 | 2 | 1 |
| 0.1 | 1.2406 | 5.7 | 4.5947 | 0.1 | 2.4 | 2.4 | 1 |
| 0.05 | 1.2718 | 7.8 | 6.1329 | 0.05 | 2.693 | 2.7 | 1.0026 |
| 0.02 | 1.2840 | 10.2 | 7.9437 | 0.02 | 2.9 | 2.9 | 1 |
| 0 ($\tau_1 = \infty$) | 1.3085 | 12 | 9.1708 | 0 ($\tau_1 = \infty$) | 3.0213 | 3.1 | 1.026 |

In addition, for lag-dominant plants, Table 8.1 shows that the value of $\lambda$ is kept fairly constant in the interval $[1.1887\theta, 1.3085\theta]$, and that is the reason why we propose selecting

$$\lambda = 1.25\theta \tag{8.9}$$

in order to achieve a robustness level of $M_S \approx 1.6$ for plants with $\theta/\tau_1 < 1$. For plants with $\theta/\tau \geq 1$, this choice is rather conservative, and it would be better to take $\lambda \approx \theta$. Note that, for the integrating case, $\gamma$ should be selected as $\gamma \approx 9\lambda$. With this choice of $\gamma$,

$$K_c = \frac{10\lambda + \theta}{k'(\lambda + \theta)(9\lambda + \theta)} \tag{8.10}$$

$$T_i = 10\lambda + \theta$$

and selecting $\lambda = 1.25\theta$ as recommended, the tuning rule for the IPTD model (7.19) reads

$$K_c \approx \frac{1}{k'}\frac{0.5}{\theta} \tag{8.11}$$

$$T_i = 13.5\theta$$

### 8.3.1.2    Tuning based on $J_{\text{avg}}$

Now, it is time to see what happens when $J = J_{\text{avg}}$. Quite surprisingly, Table 8.1 shows that using $J_{\text{avg}}$ results in a regulatory-type tuning ($\gamma \approx \lambda$). This outcome was already observed in [3], and it also agrees with the discussion in [31, Section 3.4]. Therefore, according to the $J_{\text{avg}}$ criterion, the tuning of $\gamma$ is easy; one should simply go for regulatory control, or

$$\gamma = \lambda \tag{8.12}$$

As for the value of $\lambda$, we notice from Table 8.1 that it is comprised between $2\theta$ and $3.1\theta$ for $\theta/\tau_1 \leq 0.2$. Henceforth, to make it simple, we choose

$$\lambda = 2.5\theta \tag{8.13}$$

In the integrating case, substituting (8.12) and (8.13) into (7.19), we get the following expressions for the tuning parameters

$$K_c \approx \frac{1}{k'}\frac{0.5}{\theta} \tag{8.14}$$

$$T_i = 6\theta$$

### 8.3.2    Unstable case

The robustness/performance trade-off in the unstable plant case ($\tau_1 < 0$) is shown in Figure 8.4. Compared with the stable case, the servo performance degradation ($\Delta_s^*$) is considerably higher. Also, the gap between servo ($\Delta_s^*$) and regulatory ($\Delta_r^*$) performance gets reduced in general.

As for $J_{\text{max}}^*$ and $J_{\text{avg}}^*$, the main difference with respect to the stable case, see Figure 8.3, is that one cannot specify a constant target robustness level. For example, in order to get a well-balanced robustness/performance trade-off, $M_S^t$ should be selected around 1.6 for integrating plants, $\approx 1.8$ for $\theta/\tau_1 = -0.1$, $\approx 2.2$ for $\theta/\tau_1 = -0.2$ and $\approx 2.9$ for $\theta/\tau_1 = -0.33$ (see the circles in Figure 8.4). Thus, for unstable plants, the larger the value of $\theta/|\tau_1|$, the larger the value of $M_S^t$.

### 8.3.2.1    Tuning based on $J_{\text{max}}$ and $J_{\text{avg}}$

The solutions to problem (8.6) for different values of the normalized dead time and $J = J_{\text{max}}, J_{\text{avg}}$ have been collected in Table 8.2. Contrary to what happened in the stable

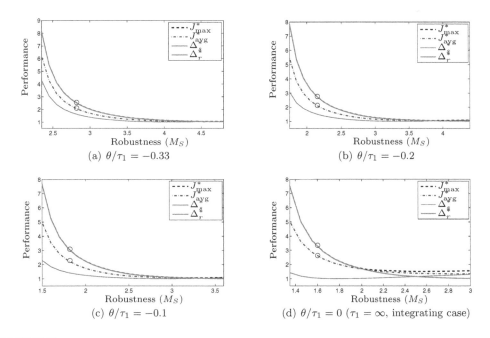

**FIGURE 8.4**
Robustness versus performance for unstable FOPTD systems (pareto-optimal curves in the $\lambda\gamma$-space).

plant case, now both $J_{\max}$ and $J_{\text{avg}}$ result in the same tuning guidelines. According to Table 8.2, $\lambda$ and $\gamma$ could be chosen as follows

$$\lambda = 3\theta \qquad \gamma = \lambda \tag{8.15}$$

The interesting point here is that $\gamma = \lambda$ regardless of the performance index used. This fact highlights that, for unstable processes, one should always go for regulatory control. To help understand this decision, let us consider the resulting closed-loop when using the controller given in (7.17). One has that

$$T = 1 - \rho W^{-1} \approx \frac{e^{-\theta s}}{\lambda s + 1}\left(\frac{\zeta s + 1}{\gamma s + 1}\right) \tag{8.16}$$

where $\zeta = T_i = \frac{\tau_1(\lambda+\gamma+\theta)-\lambda^2}{\tau_1+\theta}$. In the stable case ($\tau_1 > 0$), if $\gamma = \tau_1$ the controller cancels the pole of the plant. In this case, we have that $\zeta = \tau_1$ and, correspondingly, $T \approx \frac{e^{-\theta s}}{\lambda s+1}$, which yield good results for servo purposes. In the unstable case ($\tau_1 < 0$), however, such a pole-zero cancellation between the plant and the controller is not allowed, since cancelling the

**TABLE 8.2**
Optimal values of $\lambda, \gamma$ for problem (8.6) in the unstable FOPTD case.

| | | $J = J_{\max}$ | | | $J = J_{\text{avg}}$ | | |
|---|---|---|---|---|---|---|---|
| $\theta/\tau_1$ | $k$ | $\lambda/\theta$ | $\gamma/\theta$ | $\gamma/\lambda$ | $\lambda/\theta$ | $\gamma/\theta$ | $\gamma/\lambda$ |
| $-0.33$ | 2.9 | 3 | 3 | 1 | 3 | 3 | 1 |
| $-0.2$ | 2.2 | 2.9 | 2.9 | 1 | 2.9 | 2.9 | 1 |
| $-0.1$ | 1.8 | 3.1 | 3.1 | 1 | 3.1 | 3.1 | 1 |
| $-0.05$ | 1.7 | 3 | 3 | 1 | 3 | 3 | 1 |
| $-0.02$ | 1.6 | 3.2 | 3.3 | 1.03 | 3.2 | 3.2 | 1 |

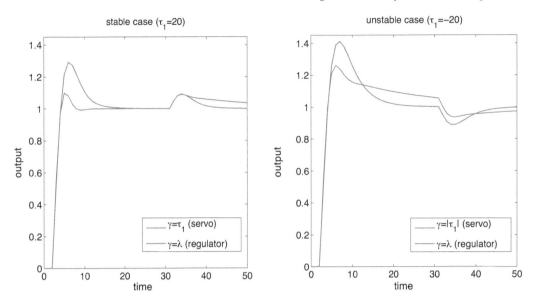

**FIGURE 8.5**
Set-point $(r)$ and load disturbance $(d_i)$ responses for $P = \frac{e^{-s}}{\tau_1 s + 1}$ and $M_S = 1.7$.

unstable pole of the plant would render the system internally unstable. Therefore, even if one chooses to tune the controller in servo mode by selecting $\gamma = |\tau_1|$, the term $\frac{\zeta s + 1}{\gamma s + 1}$ remains in the closed-loop transfer function. In addition, because $\left\| \frac{\zeta s + 1}{\gamma s + 1} \right\|_\infty > 1$, this term will produce overshoot in the set-point response (as a matter of fact, this is the case for any closed-loop system involving an unstable plant [48]). In our case, the overshoot is minimized by selecting $\gamma = |\tau_1|$, corresponding to the servo-type tuning. Nevertheless, one cannot ignore that the pole at $s = -1/\gamma = -1/|\tau_1|$ will then dominate the transient response for large values of $|\tau_1|$, resulting into a long settling time for lag-dominated processes. In practice, it turns out that the reduction in the overshoot is not worth the increased settling time (at least from an IAE point of view), and one would normally prefer a regulatory-type controller due to its faster recovery time after the inevitable overshoot. We have illustrated this point in Figure 8.5. As can be seen, in the unstable case the settling time is very long when the controller is tuned in servo mode. In particular, the process has not settled out completely when the disturbance enters at $t = 30$ sec.

For plants with a more balanced lead/lag ratio, the servo/regulation trade-off quickly dissapears, and both the servo and regulator designs exhibit large overshoots in the set-point response due to the fundamental constraint (7.27). The overall conclusion is that one should clearly go for regulatory control when dealing with unstable plants.

Substituting (8.15) into (7.17), the following settings result in

$$
\begin{aligned}
K_c &= \frac{7\tau_1 - 9\theta}{k16\theta} \\
T_i &= \frac{7\tau_1 - 9\theta}{\frac{\tau_1}{\theta} + 1}
\end{aligned}
\tag{8.17}
$$

Figure 8.6 displays the value of $M_S$ as a function of the normalized dead time for the above settings (8.17). Due to the large values of $M_S$, one should limit the use of (8.17) to plants with $\theta/|\tau_1| < 0.5$ approximately.

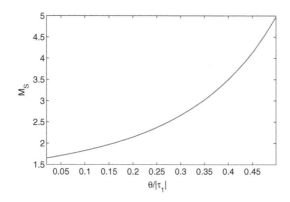

**FIGURE 8.6**
Sensitivity peak for the settings (8.17) with an unstable FOPTD process.

## 8.4 PID control using second-order models

In this section, we repeat the analysis carried out in Section 8.3, but this time we concentrate on the PID tuning rules (7.22) and (7.26), derived using SOPTD and SODIPTD models, respectively. Recall that, in (7.22), it was assumed that the model had a double real pole, that is:

$$P = \frac{ke^{-\theta s}}{(\tau_1 s + 1)^2} \qquad (8.18)$$

This case is particulary simple to deal with because, as in the FOPTD case, the analysis only depends on the normalized dead time $\theta/\tau_1$. If two different time constants, i.e., $\tau_1, \tau_2$, were considered, an extra parameter should be taken into account, namely the ratio between $\tau_1$ and $\tau_2$. In addition, in the unstable case, two cases would have to be considered: $\tau_1 < 0, \tau_2 > 0$ and $\tau_1 < 0, \tau_2 < 0$, making the analysis much more involved. Therefore, for brevity's sake, we will just consider the double-pole case. Nevertheless, since the results mainly depend on the dominant time constant, the resulting tuning guidelines can also be applied to the general SOPTD model for which $\tau_1 \neq \tau_2$.

### 8.4.1 Stable and integrating cases

As it was done in Section 8.3, we start by considering the robustness/performance trade-off to see if we can specify a constant target robustness level. In view of Figure 8.7, we select $M_S^t = 1.6$; this value is the same as it was chosen for FOPTD systems in Section 8.3.1, and it again seems to yield a good balance between robustness and performance. The next step is to study how to select the $\lambda$ and $\gamma$ parameters by solving (8.6) for $k = M_S^t = 1.6$. The resulting values for $\lambda$ and $\gamma$ have been collected in Table 8.3.

#### 8.4.1.1 Tuning based on $J_{\max}$

Here, we consider the solutions to (8.6) for $J = J_{\max}$ in Table 8.3. As for the $\gamma/\lambda$ ratio, a good fit of the data is achieved by selecting $\gamma/\lambda = \frac{1.05}{\theta/\tau_1 + 0.15}$. Because $\gamma \in [\lambda, \tau_1]$, the final selection of $\gamma$ is

$$\gamma = \min\left(\frac{1.05}{\frac{\theta}{\tau} + 0.15}\lambda, \tau_1\right) \qquad (8.19)$$

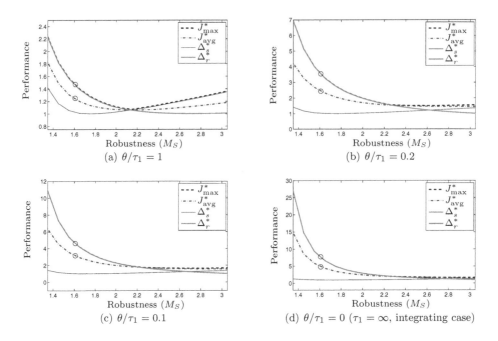

**FIGURE 8.7**
Robustness versus performance for the stable SOPTD system (8.18) (pareto-optimal curves in the $\lambda\gamma$-space).

Regarding $\lambda$, values around $\theta$ and $1.5\theta$ should be selected for balanced lead/lag ratios and lag-dominant plants ($0.02 \leq \theta/\tau_1 \leq 0.2$), respectively; and around $1.9\theta$ for the double integrating model. For the latter case, we can substitute $\gamma = 6.5256\lambda$ and $\lambda = 1.8542\theta$ (directly from Table 8.3) into (7.26) to get the following autotuning rule

$$
\begin{aligned}
K_c &= \frac{0.055}{k''\theta^2} \\
T_i &= 27\theta \\
T_d &= 8\theta
\end{aligned}
\tag{8.20}
$$

#### 8.4.1.2 Tuning based on $J_{\text{avg}}$

From Table 8.3, the tuning of $\gamma$ based on the $J_{\text{avg}}$ criterion is easy:

$$
\gamma = \lambda
\tag{8.21}
$$

**TABLE 8.3**
Optimal values of $\lambda, \gamma$ for problem (8.6), taking $k = 1.6$, in the stable SOPTD case.

| | $J = J_{\text{max}}$ | | | | $J = J_{\text{avg}}$ | | |
|---|---|---|---|---|---|---|---|
| $\theta/\tau_1$ | $\lambda/\theta$ | $\gamma/\theta$ | $\gamma/\lambda$ | $\theta/\tau_1$ | $\lambda/\theta$ | $\gamma/\theta$ | $\gamma/\lambda$ |
| 1 | 1 | 1 | 1 | 1 | 1 | 1 | 1 |
| 0.2 | 1.347 | 3.9 | 2.8954 | 0.2 | 2.65 | 2.65 | 1 |
| 0.1 | 1.4594 | 6.2 | 4.2483 | 0.1 | 3.3957 | 3.5 | 1.0307 |
| 0.05 | 1.5755 | 8.5 | 5.3952 | 0.05 | 4.15 | 4.15 | 1 |
| 0.02 | 1.6747 | 10.8 | 6.4488 | 0.02 | 4.72 | 4.72 | 1 |
| $0 \ (\tau = \infty)$ | 1.8542 | 12.1 | 6.5256 | $0 \ (\tau = \infty)$ | 5.2 | 5.2 | 1 |

Thus, as in the FOPTD case, the use of the $J_{\mathrm{avg}}$ index results in a regulatory-type tuning. The tuning of $\lambda$ is not so simple though, and one cannot specify a constant value. A good fit of the experimental data is achieved by selecting

$$\lambda = \left( \frac{0.2\frac{\theta}{\tau_1} + 1}{\frac{\theta}{\tau_1} + 0.2} \right) \theta \tag{8.22}$$

For the double-integrating case, we can substitute $\gamma = \lambda = 5.2\theta$ (directly from Table 8.3) into (7.26), which (approximately) results in

$$
\begin{aligned}
K_c &= \frac{0.07}{k''\theta^2} \\
T_i &= 16.6\theta \\
T_d &= 5.9\theta
\end{aligned}
\tag{8.23}
$$

### 8.4.2 Unstable case

Figure 8.8 shows the robustness/performance trade-off in the unstable plant case. As it happened for unstable FOPTD systems, $M_S^t$ should be greater the greater the value of $\theta/|\tau_1|$. This is because, for unstable plants, stability problems arise as we increase the normalized dead time.

#### 8.4.2.1 Tuning based on $J_{\mathrm{max}}$ and $J_{\mathrm{avg}}$

In accordance with Figure 8.8, we have solved problem (8.6) for $\theta/\tau_1 = -0.02, -0.05, -0.1$ taking $k = M_S^t = 1.7, 1.8, 2.2$, respectively. Both for $J_{\mathrm{max}}$ and $J_{\mathrm{avg}}$,

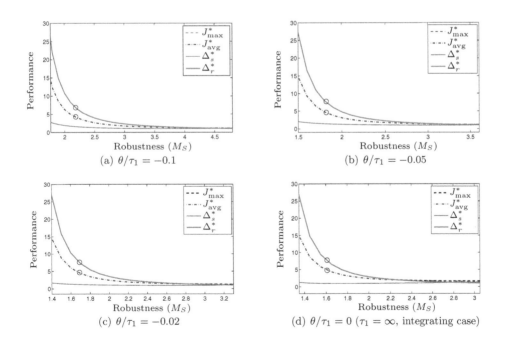

**FIGURE 8.8**

Robustness *vs* performance for the unstable SOPTD system (8.18) (pareto-optimal curves in the $\lambda\gamma$-space).

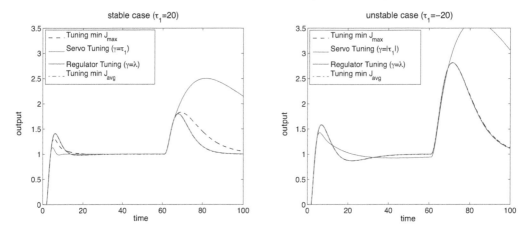

**FIGURE 8.9**
Set-point $(r)$ and load disturbance $(d_i)$ responses for $P = \frac{50e^{-s}}{(\tau_1 s+1)^2}$ and $M_S = 1.8$.

the overall conclusion is to select $\gamma \approx \lambda$, that is, to go for regulatory control. This fact obeys the same reason as in the unstable FOPTD case: tuning the controller in servo mode $(\gamma = |\tau_1|)$ reduces the overshoot in the set-point response at the expense of an increased settling time. Then, because of the compromise between overshoot and settling time, the set-point response is not really improved compared to that achieved in regulator mode $(\gamma = \lambda)$. However, the disturbance response for the regulator mode is much better than for the servo one, making the decision easy, see Figure 8.9.

From the conducted experiments, it has been observed that $\gamma \approx 5.2\theta$. In summary, the tuning guideline for the unstable SOPTD case is

$$\gamma = \lambda = 5.2\theta \tag{8.24}$$

Figure 8.10 displays the value of $M_S$ for different values of the normalized dead time using the settings (7.20) together with the above recommendation (8.24). Due to the large values of $M_S$, one should limit the applicability to $\theta/|\tau_1| \leq 0.2$ approximately.

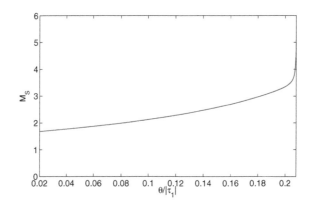

**FIGURE 8.10**
Sensitivity peak for the settings (7.20) with $\lambda = \gamma = 5.2\theta$ and $P = \frac{ke^{-\theta s}}{(\tau_1 s+1)^2}, \tau_1 < 0$.

# A

## Appendix

This appendix contains the ISE calculations of Sections 4.2 and 4.3.

### ISE calculation of Section 4.2

From the first-order approximation of the FOPTD model, we have from (5.31), (3.15), and (3.25) that

$$y = (1 - P)d_o = \left(1 - \frac{(-hs + 1)((0.2426\gamma + \sqrt{2}h)s + 1)}{(1 + \sqrt{2}hs)(1 + 2hs)}\right)\frac{1}{s} \qquad (A.1)$$

The $\gamma$ parameter can be fixed according to the ISE criterion

$$ISE(\gamma, h) = \int_0^\infty y^2(t)dt = \|y\|_2^2 \qquad (A.2)$$

By applying the Parseval's theorem [49], this calculation can be rewritten as

$$ISE(\gamma, h) = \frac{1}{2\pi}\int_{-\infty}^\infty y(j\omega)y(-j\omega)d\omega = \frac{1}{2\pi j}\oint y(s)y(-s)ds \qquad (A.3)$$

The last integral is a contour integral up the imaginary axis, then an infinite semicircle in the left half-plane. The contribution from this semicircle is zero because the integrand is strictly proper. By the residue theorem [24, 49], the above circulation integral equals the sum of the residues of $y(s)y(-s)$ at its poles in the left half-plane. More precisely,

$$
\begin{aligned}
ISE(\gamma, h) &= \mathrm{Res}\left(y(s)y(-s), -\frac{1}{\sqrt{2}h}\right) + \mathrm{Res}\left(y(s)y(-s), -\frac{1}{2h}\right) \\
&= 2.25h - 0.1065\gamma + \frac{0.0112}{h}\gamma^2 \qquad (A.4)
\end{aligned}
$$

By taking the derivative with respect to $\gamma$, we can obtain the optimal value that minimizes the ISE criterion

$$\frac{\partial ISE}{\partial \gamma} = 0 \Rightarrow \gamma_{ldo} \approx 4.56h \qquad (A.5)$$

producing the following value for the optimum $ISE$

$$ISE_{op}(h) \approx 1.99h \qquad (A.6)$$

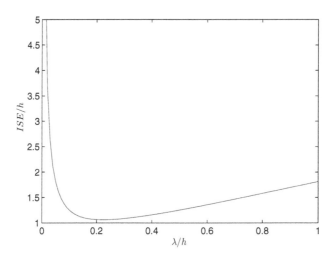

**FIGURE A.1**
$ISE/h$ versus $\lambda/h$.

## ISE calculation of Section 4.3

From (2.5) and (2.37), we have that

$$y = (1 - PQ)d_o = \left(1 - \frac{-\frac{h}{2}s + 1}{(\lambda s + 1)^2}\right)\frac{1}{s} = \frac{\lambda^2 s + 2\lambda + \frac{h}{2}}{(\lambda s + 1)^2} \tag{A.7}$$

Along the lines of Section 4.2, the ISE criterion can be calculated as the sum of the residues of $y(s)y(-s)$ at its poles in the left half-plane (LHP):

$$ISE(\lambda, h) = \text{Res}\left(y(s)y(-s), -\frac{1}{\lambda}\right) = h\frac{20\left(\frac{\lambda}{h}\right)^2 + 1 + 8\frac{\lambda}{h}}{16\frac{\lambda}{h}} \tag{A.8}$$

The plot of $ISE/h$ versus $\lambda/h$ appears in Figure A.1. Now, by taking the derivative with respect to $\lambda$, we can obtain the value that minimizes the ISE criterion

$$\frac{\partial ISE}{\partial \lambda} = 0 \Rightarrow \lambda = \frac{\sqrt{5}h}{10} \approx 0.22h \tag{A.9}$$

producing the following value

$$ISE_{op}(h) \approx 1.06h \tag{A.10}$$

Robustness associated with the tuning $\lambda = 0.22h$ is poor ($M_S$ can be seen to be around 3). As commented in Section 4.3, for *tight* control the idea is to increase $\lambda$ so as to achieve acceptable robustness margins. This will provide the fastest-possible closed-loop system since $\lambda$ directly controls the closed-loop bandwidth. In concrete, this implies that the best disturbance attenuation will be obtained. Although only output disturbance has been explicitly considered, it is easy to check that the smaller the value of $\lambda$, the better the load disturbance rejection as well. To see this, let us consider the output to a load disturbance[1]:

$$y = P(1 - PQ)d_i = \frac{(-\frac{h}{2}s + 1)(\lambda^2 s + 2\lambda + \frac{h}{2})}{(\frac{h}{2}s + 1)(\tau s + 1)(\lambda s + 1)^2} \tag{A.11}$$

---

[1]In what follows, we assume $K_g = 1$ in (2.5).

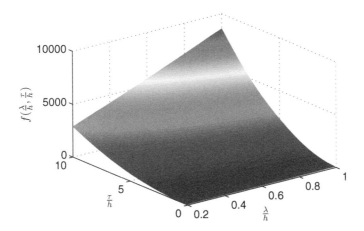

**FIGURE A.2**
$f$ versus $\lambda/h$ and $\tau/h$.

Proceeding as before, the associated ISE can be computed as the sum of the residues of $y(s)y(-s)$ in its LHP poles:

$$ISE(\lambda, \tau, h) = \frac{1}{4} \frac{5\lambda^3 + 8\tau\lambda^2 + 2\lambda^2 h + \frac{1}{4}h^2\lambda + 4\tau\lambda + \frac{1}{2}\tau h^2}{(\lambda + \tau)^2} \qquad (A.12)$$

Straightforward calculations yield

$$\frac{\partial ISE(\lambda, \tau, h)}{\partial \lambda} = \frac{20\lambda^3 + 60\tau\lambda^2 - h^2\lambda + 64\tau^2\lambda - 3h^2\tau + 16h\tau^2}{16(\lambda + \tau)^3} \qquad (A.13)$$

The sign of $\frac{\partial ISE}{\partial \lambda}$ is determined by the sign of the numerator in (A.13). Dividing it by $h^3$, the sign can be determined in terms of the following function:

$$f(\frac{\lambda}{h}, \frac{\tau}{h}) = 20\left(\frac{\lambda}{h}\right)^3 + 60\left(\frac{\lambda}{h}\right)^2\frac{\tau}{h} - \frac{\lambda}{h} + 64\left(\frac{\tau}{h}\right)^2\frac{\lambda}{h} - 3\frac{\tau}{h} + 16\left(\frac{\tau}{h}\right)^2 \qquad (A.14)$$

The plot of $f$ is shown in Figure A.2.

As can be appreciated, $f$ remains positive. This means that the ISE with respect to load disturbance is an increasing function of $\lambda$ in the considered interval $(0.22 < \frac{\lambda}{h} \leq 1)$.

# Bibliography

[1] L. A. Aguirre. PID tuning based on model matching. *Electronic Letters*, 28(25): 2269–2271, 1992.

[2] S. Alcántara, C. Pedret, and R. Vilanova. On the model matching approach to PID design: Analytical perspective for robust servo/regulation trade-off tuning. *Journal of Process Control*, 20(5):596–608, 2010.

[3] S. Alcántara, R. Vilanova, C. Pedret, and S. Skogestad. A look into robustness/ performance and servo/regulation issues in PI tuning. In *Proc. of the IFAC Conf. on Advances in PID Control PID'12*, Brescia, Italy, pp. 181–186, 2012.

[4] S. Alcántara, C. Pedret, and R. Vilanova. On the model matching approach to PID design: Analytical perspective for robust servo/regulation trade-off tuning. *Journal of Process Control*, 20(5):596–608, 2010.

[5] S. Alcántara, C. Pedret, R. Vilanova, and P. Balaguer. Analytical $\mathcal{H}_\infty$ sensitivity matching approach to smooth PID design: quantitative servo/regulation tuning guidelines. Internal report, 2010.

[6] S. Alcántara, C. Pedret, R. Vilanova, and S. Skogestad. Generalized internal model control for balancing input/output disturbance response. *Industrial & Engineering Chemistry Research*, 50(19):11170–11180, 2011.

[7] S. Alcántara, C. Pedret, R. Vilanova, and W. Zhang. Analytical $\mathcal{H}_\infty$ design for a Smith-type inverse response compensator. In *Proc. of the American Control Conference*, St. Louis, MO, USA, pp. 1604–1609, 2009.

[8] S. Alcántara, C. Pedret, R. Vilanova, and W. Zhang. Simple analytical min-max model matching approach to robust proportional-integrative-derivative tuning with smooth set-point response. *Industrial & Engineering Chemistry Research*, 49(2):690–700, 2010.

[9] S. Alcántara, S. Skogestad, C. Grimholt, C. Pedret, and R. Vilanova. Tuning PI controllers based on $\mathcal{H}_\infty$ weighted sensitivity. In *19th Mediterranean Conference on Control and Automation (MED 2011)*, pp. 1301–1306, Corfu, Greece, June 2011.

[10] S. Alcántara, W.D. Zhang, C. Pedret, R. Vilanova, and S. Skogestad. IMC-like analytical $\mathcal{H}_\infty$ design with S/SP mixed sensitivity consideration: utility in PID tuning guidance. *Journal of Process Control*, 21(6):976–985, 2011.

[11] V. Alfaro and R. Vilanova. Conversion formulae and performance capabilities of two-degree-of-freedom PID control algorithms. In *Proc. of the 17th IEEE International Conference on Emerging Technologies & Factory Automation (ETFA' 2012)*, Krakow, Poland, pp. 1–6, 2012.

[12] V. M. Alfaro, R. Vilanova, R. Méndez, and J. Lafuente. Performance/robustness trade-off analysis of PI/PID servo and regulatory control systems. In *Proc. of the IEEE International Conference on Industrial Technology*, Vina del Mar, CHile, pp. 111–116, 2010.

[13] A. Ali and S. Majhi. PI/PID controller design based on IMC and percentage overshoot specification to controller set-point change. *ISA Transactions*, 48:10–15, 2009.

[14] M. S. Amiri and S. L. Shah. Guidelines on robust PID controller tuning for FOPTD processes. In *Proc. of the 8th World Congress of Chemical Engineering*, Montreal, PQ, August 2009.

[15] M. Araki and H. Taguchi. Two-degree-of-freedom PID controllers. *International Journal of Control, Automation, and Systems*, 1(4):401–411, 2003.

[16] O. Arrieta and R. Vilanova. Servo/regulation trade-off tuning of PID controllers with a robustness consideration. *Proc of the 46th Conference on Decision and Control*, New Orleans, LA, pp. 1838–1843, 2007.

[17] O. Arrieta, A. Visioli, and R. Vilanova. PID autotuning for weighted servo/regulation control operation. *Journal of Process Control*, 20(4):472–480, 2010.

[18] K. Astrom and T. Hagglund. The future of PID control. *Control Engineering Practice*, 9(11):1163–1175, 2001.

[19] K. Astrom and T. Hagglund. Revisiting the Ziegler-Nichols step response method for PID control. *J. Process Control*, 14:635–650, 2004.

[20] K. Astrom and T. Hagglund. *Advanced PID control.* ISA—The Instrumentation, Systems, and Automation Society, 2005.

[21] P. Balaguer, A. Ibeas, C. Pedret, and S. Alcántara. Controller parameters dependence on model information through dimensional analysis. *Proceedings of the 48h IEEE Conference on Decision and Control (CDC) held jointly with 2009 28th Chinese Control Conference*, Shanghai, 2009, pp. 1914–1919.

[22] M. Campi, W. S. Lee, and B. D. O. Anderson. New filters for internal model control design. *International Journal of Robust and Nonlinear Control*, 4(6):757–775, 1994.

[23] I. L. Chien and P. S. Fruehauf. Consider IMC tuning to improve controller performance. *Chemical Engineering Progress*, 86(10):33–41, 1990.

[24] R. Churchill and J. Brown. *Complex Variable and Applications.* New York: McGraw-Hill, 1986.

[25] A. Dehghani, A. Lanzon, and B.O. Anderson. $\mathcal{H}_\infty$ design to generalize internal model control. *Automatica*, 42(11):1959–1968, 2006.

[26] J.C. Doyle, B.A. Francis, and A. Tanenbaum. *Feedback Control Theory.* New York: MacMillan Publishing Company, 1992.

[27] A. Faanes and S. Skogestad. Feedforward control under the presence of uncertainty. *European Journal of Control*, 10(1):30–46, 2004.

[28] B. A. Francis. *A Course in $\mathcal{H}_\infty$ Control Theory.* Springer-Verlag, Berlin. Lecture Notes in Control and Information Sciences, 1987.

[29] O. Garpinger, T. Hagglund, and K. Astrom. Criteria and trade-offs in PID design. In *Proc. of the IFAC Conf. on Advances in PID Control PID'12*, Brescia, Italy, pp. 47–52, 2012.

[30] C. Grimholt [Sigurd Skogestad]. Should we forget the smith predictor? 3rd IFAC Conference on Advances in Proportional-Integral-Derivative Control PID 2018. *IFAC-PapersOnLine*, 51(4):769–774, 2018.

[31] C. Grimholt and S. Skogestad. Optimal PI control and verifcation of the SIMC tuning rule. In *Proc. of the IFAC Conf. on Advances in PID Control PID'12*, Brescia, Italy, pp. 11–22, 2012.

[32] J. He, Q. Wang, and T. Lee. PI/PID controller tuning via LQR approach. *Chemical Engineering Science*, 55:2429–2439, 2000.

[33] N. Hohenbichler. All stabilizing PID controllers for time delay systems. *Automatica*, 45(11):2678–2684, 2009.

[34] I. G. Horn, J. R. Arulandu, C. J. Gombas, J. G. VanAntwerp, and R. D. Braatz. Improved filter design in internal model control. *Industrial & Engineering Chemistry Research*, 35(10):3437–3441, 1996.

[35] A. Ibeas and S. Alcántara. Stable genetic adaptive controllers for multivariable systems using a two-degree-of-freedom topology. *Engineering Applications of Artificial Intelligence*, 23(1):41–47, 2010.

[36] A. Isaakson and S. Graebe. Derivative filter is an integral part of PID design. *IEE Proc. Part D*, 149:41–45, 2002.

[37] M. Kano and M. Ogawa. The state of the art in chemical process control in Japan: Good practice and questionnaire survey. *Journal of Process Control*, 20(9):969–982, 2010.

[38] B. Kristiansson and B. Lennartson. Optimal PID controllers for unstable and resonant plants. In *Proceedings of the 37th IEEE Conference on Decision and Control (Cat. No.98CH36171)*, Tampa, FL, USA, 1998, pp. 4380–4381.

[39] B. Kristiansson and B. Lennartson. Robust tuning of PI and PID controllers: using derivative action despite sensor noise. *IEEE Control Systems Magazine*, 26(1):55–69, 2006.

[40] W.W. Kwok and D.E. Davison. Implementation of stabilizing control laws—how many controller blocks are needed for a universally good implementation? *Control Systems Magazine, IEEE*, 27(1):55–60, 2007.

[41] Per-Ola Larsson and Tore Hagglund. Control signal constraints and filter order selection for PI and PID controllers. In *Proceedings of the 2011 American Control Conference*, San Francisco, CA, 2011, pp. 4994–4999.

[42] W. S. Lee and J. Shi. Improving $\mathcal{H}_\infty$ to generalize internal model control with integral action. In *2008 Chinese Control and Decision Conference*, Yantai, Shandong, 2008, pp. 4144–4149.

[43] Y. Lee, J. Lee, and S. Park. PID controller tuning for integrating and unstable processes with time delay. *Chemical Engineering Science*, 55(17):3481–3493, 2000.

[44] A. Leva and M. Maggio. A systematic way to extend ideal PID tuning rules to the real structure. *Journal of Process Control*, 21(1):130–136, 2011.

[45] A. Leva and M. Maggio. Model-based PI(D) autotuning. In: Vilanova R., Visioli A. (eds) *PID Control in the Third Millennium*, pp. 45–73. Advances in Industrial Control. Springer, London, 2012.

[46] W. Luyben. Effect of derivative algorithm and tuning selection on the PID control of dead-time processes. *Industrial & Engineering Chemistry Research*, 40:3605–3611, 2001.

[47] D. C. McFarlane and K. Glover. A loop shaping design procedure using $\mathcal{H}_\infty$ synthesis. *IEEE Trans. Automat. Contr.*, 37(6):759–769, 1992.

[48] Richard H. Middleton and Stefan F. Graebe. Slow stable open-loop poles: to cancel or not to cancel. *Automatica*, 35(5):877–886, 1999.

[49] M. Morari and E. Zafiriou. *Robust Process Control*. Prentice-Hall International, Englewood Cliffs, N.J.: Prentice Hall, 1989.

[50] J. E. Normey-Rico and J. L. Guzmán. Unified PID tuning approach for stable, integrative and unstable dead-time processes. In *Proc. of the IFAC Conf. on Advances in PID Control PID'12*, Brescia (Italy), pp. 35–40, 2012.

[51] A. O'Dwyer. *Handbook of PI and PID controller tuning rules*. Imperial College Press, 2nd ed., Covent Garden, London, 2006.

[52] B. Ogunnaike and K. Mukati. An alternative structure for next generation regulatory controllers: Part I: Basic theory for design, development and implementation. *Journal of Process Control*, 16(5):499–509, 2006.

[53] L. Ou, W. Zhang, and L. Yu. Low-order stabilization of LTI systems with time delay. *IEEE Trans. Automat. Contr.*, 54(4):774–787, 2009.

[54] F. Padula and A. Visioli. Tuning rules for optimal PID and fractional-order PID controllers. *Journal of Process Control*, 21(1):69–81, 2011.

[55] H. Panagopoulos and K. J. Astrom. PID control design and $\mathcal{H}_\infty$ loop shaping. *Int. J. Robust Nonlinear Control*, 10:1249–1261, 2000.

[56] Gabriele Pannocchia, Nabil Laachi, and James B. Rawlings. A candidate to replace PID control: SISO-constrained LQ control. *AIChE Journal*, 51(4):1178–1189, 2005.

[57] C. Pedret, R. Vilanova, R. Moreno, and I. Serra. A refinement procedure for PID controller tuning. *Computers & Chemical Engineering*, 26(6):903–908, 2002.

[58] L. Pernebo. An Algebraic Theory for the Design of Controllers for Linear Multivariable Systems-Part II: Feedback Realizations and Feedback Design. *IEEE Transactions on Automatic Control*, 26(1):183–194, 1981.

[59] A. Seshagiri Rao and M. Chidambaram. PI/PID controllers design for integrating and unstable systems. In: Vilanova R. and Visioli A. (eds.) *PID Control in the Third Millennium*, pp. 75–111, Advances in Industrial Control. Springer, London, 2012.

[60] Khalid El Rifai. Nonlinearly parameterized adaptive pid control for parallel and series realizations. In *Proceedings of the 2009 American Control Conference*, St. Louis, MO, 2009, pp. 5150–5155.

[61] Daniel E. Rivera, Manfred Morari, and Sigurd Skogestad. Internal model control: PID controller design. *Industrial & Engineering Chemistry Process Design and Development*, 25(1):252–265, 1986.

[62] R. Sanchís, J. A. Romero, and P. Balaguer. Tuning of PID controllers based on simplified single parameter optimisation. *International Journal of Control*, 83(9):1785–1798, 2010.

[63] C. Scali and D. Semino. Performance of optimal and standard controllers for disturbance rejection in industrial processes. In *Proceedings IECON '91: 1991 International Conference on Industrial Electronics, Control and Instrumentation*, Kobe, Japan, 1991, pp. 2033–2038.

[64] M. Shamsuzzoha and M. Lee. IMC-PID controller design for improved disturbance rejection of time-delayed processes. *Industrial & Engineering Chemistry Research*, 46(7):2077–2091, 2007.

[65] M. Shamsuzzoha and M. Lee. Enhanced disturbance rejection for open-loop unstable process with time delay. *ISA Transactions*, 48(2):237–244, 2009.

[66] M. Shamsuzzohaa and S. Skogestad. The set-point overshoot method: A simple and fast closed-loop approach for PID tuning. *Journal of Process Control*, 20(10):1220–1234, 2010.

[67] J. Shi and W.S. Lee. Set-point response and disturbance rejection trade-off for second-order plus dead time processes. In *2004 5th Asian Control Conference (IEEE Cat. No.04EX904)*, Melbourne, Victoria, Australia, 2004, pp. 881–887, Vol. 2.

[68] G. J. Silva, A. Datta, and S. P. Bhattacharyya. New results on the synthesis of pid controllers. *IEEE Transactions On Automatic Control*, 47(2):241–252, 2002.

[69] S. Skogestad. Simple analytic rules for model reduction and PID controller tuning. *J. Process Control*, 13:291–309, 2003.

[70] S. Skogestad. Tuning for smooth PID control with acceptable disturbance rejection. *Industrial & Engineering Chemistry Research*, 45:7817–7822, 2006.

[71] S. Skogestad and C. Grimholt. PID tuning for smooth control. In: Vilanova R. and Visioli A. (eds.) *PID Control in the Third Millennium*, pp. 147–175, Advances in Industrial Control. Springer, London, 2012.

[72] S. Skogestad and I. Postlethwaite. *Multivariable Feedback Control*. John Wiley & Sons, Chichester | New York | Brisbane | Toronto | Singapore, 2005.

[73] Jacques F. Smuts. *Process Control for Practitioners: How to Tune PID Controllers and Optimize Control Loops*. OptiControls, 2011.

[74] Y. Songa, M. Tadñe, and T. Zhang. Stabilization and algorithm of integrator plus dead-time process using PID controller. *Journal of Process Control*, 19(9):1529–1537, 2009.

[75] G. Szita and C.K. Sanathanan. Model matching controller design for disturbance rejection. *J. Franklin Inst.*, 333(B)(5):747–772, 1996.

[76] J. Sánchez, A. Visioli, and S. Dormido. A two-degree-of-freedom PI controller based on events. *Journal of Process Control*, 21(4):639–651, 2011.

[77] S. Tavakoli, I. Griffin, and P. Fleming. Robust PI control design: a genetic algorithm approach. *Int. Journal of Soft Computing*, 2(3):401–407, 2007.

[78] R. Toscano. A simple PI/PID controller design method via numerical optimization approach. *J. Process Control*, 15:81–88, 2005.

[79] J. G. Van de Vusse. Plug-flow type reactor versus tank reactor. *Chem. Eng. Sci.*, 19:964, 1964.

[80] M. Vidyasagar. *Control System Synthesis. A factorization approach*. MIT Press. Cambridge, Massachusetts, 1985.

[81] R. Vilanova. IMC based Robust PID design: Tuning guidelines and automatic tuning. *Journal of Process Control*, 18:61–70, 2008.

[82] R. Vilanova and O. Arrieta. PID design for improved disturbance attenuation: min max Sensitivity matching approach. *IAENG International Journal of Applied Mathematics*, 37(1), 2007.

[83] R. Vilanova, O. Arrieta, and P. Ponsa. IMC based feedforward controller framework for disturbance attenuation on uncertain systems. *ISA Transactions*, 48(4):439–448, 2009.

[84] R. Vilanova and I. Serra. Realization of two-degree-of-freedom compensators. *IEE Proceedings. Part D.*, 144(6):589–596, 1997.

[85] R. Vilanova and I. Serra. Model reference control in two-degree-of-freedom control systems: Adaptive min-max approach. *IEE Proceedings. Part D.*, 146(3):273–281, 1999.

[86] A. Visioli. Optimal tuning of PID controllers for integral and unstable processes. *IEE Proceedings. Part D*, 148(2):180–184, 2001.

[87] A. Visioli. Improving the load disturbance rejection performances of IMC-tuned PID controllers. In *Proc. of the 15th IFAC Triennial World Congress*, pp. 295–300, Barcelona, Spain, 2002.
Improving the load disturbance rejection performances of IMC-tuned PID controllers. In *Proc. of the 15th IFAC Triennial World Congress*, 2002.

[88] Q-G. Wang, C. hang, and X-P. Yang. Single-loop controller design via IMC principles. *Automatica*, 37(12):2041–2048, 2001.

[89] D. Youla, H.A. Jabr, and J. Bongiorno. Modern Wiener-Hopf design of optimal controllers, part II: The multivariable case. *IEEE Trans. Automat. Contr.*, 21(6):319–338, 1976.

[90] G. Zames and B. Francis. Feedback, minimax sensitivity, and optimal robustness. *IEEE Trans. Automat. Contr.*, AC-28(5):585–601, 1983.

[91] W. Zhang, F. Allgower, and T. Liu. Controller parameterization for SISO and MIMO plants with time delay. *Systems & Control Letters*, 55(10):794–802, 2006.

[92] W. Zhang, Y. Xi, G. Yang, and X. Xu. Design PID controllers for desired time-domain or frequency-domain response. *ISA Transactions*, 41(4):511–520, 2002.

[93] Weidong Zhang. *Quantiative Process Control Theory*. Boca Raton, FL: CRC Press, 2012.

[94] Q.-C. Zhong. Robust stability analysis of simple systems controlled over communication networks. *Automatica*, 39(7):1309–1312, 2003.

[95] K. Zhou and Z. Ren. A new controller architecture for high performance, robust and fault tolerant control. *IEEE Trans. Automat. Contr.*, 46(10):1613–1618, 2001.

[96] M. Zhuang and D. Atherton. Automatic tuning of optimum PID controllers. *IEE Proceedings. Part D*, 140(3):216–224, 1993.

[97] L. G. Ziegler and N. B. Nichols. Optimum settings for automatic controllers. *ASME*, 64:759–768, 1942.

# Index